Environmental Radiochemical Analysis VII

Environmental Radiochemical Analysis VII

Edited by

Nicholas Evans
Nottingham Trent University, UK
Email: nicholas.evans@ntu.ac.uk

ROYAL SOCIETY OF CHEMISTRY

The proceedings are based on the 14th International Symposium on Nuclear and Environmental Radiochemical Analysis held in York, UK on 12-15 September 2022.

Special Publication No. 357

Print ISBN: 978-1-83767-063-5
PDF ISBN: 978-1-83767-075-8

A catalogue record for this book is available from the British Library

The Royal Society of Chemistry is a charity, registered in England and Wales, Number 207890, and a company incorporated in England by Royal Charter (Registered No. RC000524), registered office: Burlington House, Piccadilly, London W1J 0BA, UK, Telephone: +44 (0) 20 7437 8656.

Visit our website at www.rsc.org/books

Printed in the United Kingdom by CPI Group (UK) Ltd, Croydon, CR0 4YY, UK

EXHIBITORS:

ANTECH

www.antech-inc.com

BERTHOLD

TRISKEM

Expertise in Separation Chemistry

Southern Scientific

EXPERIENCE & EXPERTISE

GAU

Radioanalytical Laboratories

NNUF EXACT

University of Southampton

RADDEC INTERNATIONAL

SOCOTEC

THE POWER OF FORESIGHT

Preface

Environmental Radiochemical Analysis VII is a collection of refereed papers presented at the 14th International Symposium on Nuclear and Environmental Radiochemical Analysis (ERA14), held at The National Railway Museum, York, UK on 12–15 September 2022. This symposium is the latest in a long-running series of international symposia that has been held every four years by the Radiochemistry Group of the Royal Society of Chemistry, for over 50 years. It is the seventh to have its proceedings published as a book by the Royal Society of Chemistry. The scope of the conference has changed gradually over those years and from an initial focus on the development and application of methods for the analysis of radionuclides in environmental samples, and now includes sessions covering:

- Studies of environmental source terms, transport pathways and impacts,
- Nuclear and non-nuclear industry waste characterisation and disposal, and
- Nuclear and Environmental forensics.

Thirty-nine papers were presented as oral presentations, with an additional 43 as posters. The conference was attended by delegates representing 27 different countries, with a total attendance of 142, with a significant number of attendees from outside the UK.

This year's conference was the first since 2018 and the first major conference organised by the Radiochemistry Group since COVID-19. As such, it was a bit of a step into the unknown, in terms of number of attendees, especially with the possibility of further travel restrictions. But, the level of interest was high, with applications exceeding the number of places available. Professor Dave Mills (University of Manchester) received the Bill Newton Award (2019) and Professor Francis Livens (Dalton Nuclear Institute) received the Becquerel Medal (2021), with both giving invited lectures.

That the conference was so successful is due to a lot of organisation and hard work and I would like to thank the members of both the Organising Committee and the Technical Committee, particularly Nick Evans (Nottingham Trent University) and Dave Wickenden (University of Surrey), who carried out most of the groundwork, including getting sponsorship, securing the wonderful venue and generally making sure that things went smoothly. I would also like to thank Rich March (University of Southampton) for carrying out the financial aspects that are associated with the conference. Finally, a big thanks to the presenters, without whom there would have been no conference.

See you at ERA15, in 2026!

Mike Collins
Chair, ERA14 Conference Organising Committee

Contents

CHAPTER 1 LOCALISATION, ISOLATION AND LEACHING OF SINGLE HOT PARTICLES FROM VARIOUS ENVIRONMENTAL MATRICES IN THE VICINITY OF THE CHERNOBYL NUCLEAR POWER PLANT

Laura Leifermann[a*], Martin Weiss[a*], Ihor Chyzhevskyi[b], Sergiy Dubchak[a], Paul Hanemann[a], Manuel Raiwa[a], Wolfgang Schulz[a], Georg Steinhauser[a], Tobias Weissenborn[a] and Clemens Walther[a]

[a]Leibniz Universität Hannover, Institut für Radioökologie und Strahlenschutz (IRS), Herrenhäuser Str. 2, 30419 Hannover, Germany, [b]State Specialised Enterprise "Ecocentre" (SSE Ecocentre), 07270, Shkilna str.6, Chernobyl, Ukraine.

*leifermann@irs.uni-hannover.de

1 INTRODUCTION

Radioactive particles can be found in the environment in various regions of the world. Reactor accidents such as in Chernobyl and Fukushima have deposited microscopic particles with varying composition in the environment. Whereas in Chernobyl nuclear fuel fragments were released containing both actinides and fission products, Fukushima released glass particles containing mostly caesium isotopes.[1,2,3] Further sources of uranium- or plutonium-containing particles include the nuclear weapons explosions, either from testing or accidental destruction, as in the Palomares incident.[4]

The Chernobyl Nuclear Power Plant (ChNPP) accident on the 26th of April 1986 released roughly 1.5 ± 0.5% of the reactor fuel.[5] This was deposited as radioactive particles in the area, predominantly 0-30 km from the destroyed reactor unit four.[6,7,8] The radionuclide composition originally carried by these particles was highly variable, depending on the status of fuel burn-up at the point of origin within the reactor assembly as well as annealing period and temperature.[9] While the activity-bearing core of these particles originally consisted of partially spent uranium dioxide fuel, it was in part chemically altered during the explosion and reactor fire. These different particles exhibit distinguishable morphological characteristics and chemical properties and can roughly be categorised into three different groups.[10,11,12]

The first well-established group is represented by uranium dioxide particles (UO_2) that have not been significantly oxidised or otherwise chemically altered from the original fuel. Depending on soil characteristics (mostly pH) they have a dissolution half-life of 7 to 70 years.[12,13] They were formed by the disintegration of fuel during the initial explosion.[9] These particles exhibit a porous structure very similar to the original fuel.[11,14,15]

The second group consists of oxidised particles with an O to U ratio significantly higher than 2 ($UO_{(2+x)}$). These particles have been described as environmentally unstable with a faster dissolution half-life ranging from 1 to 7 years.[13] The high temperatures of the burning reactor and the direct contact with air from the 26th of April to the 5th of May led to the oxidation and ejection of this class of hot particles.[10,11,13,15]

The last group are the zirconium-uranium-particles ($U_XZr_YO_Z$ particles) with high stability and little to no dissolution rate under environmental conditions. This class of particles was created during the initial explosion, when the Zircalloy cladding merged with the uranium dioxide fuel. The surface of these particles is smooth and shows almost no pores.[10,12,15]

The original size of the emitted particles roughly follows a log-normal distribution with a median radius between 3 and 10 µm.[9,16,17,18] Bound within their respective uranium dioxide matrix, the particles still carry a whole range of activation and fission products.[2,3,11] It is – amongst others – mainly the presence of ^{241}Am, $^{239+240}$Pu, ^{241}Pu, ^{137}Cs, ^{90}Sr and ^{154}Eu carried within these particles that poses a present and future threat through the ongoing contamination in the near environment of the ChNPP.[19,20]

Due to its eminent importance in risk assessment for the contaminated lands, the dissolution behaviour of these particles has been subject to extensive environmental studies over the past few decades.[5,13,19,21,22] Many of these studies are based on radionuclide extraction patterns of bulk soil samples taken from spots with presumed particulate contamination, yielding an overview of the average nuclide content in soil. This well-established method is incredibly useful as it reliably accounts for even the smallest sized particles, which may otherwise be neglected.[23] In a given matrix, it even yields some information on the chemical composition of the (sequentially) leached particles. However, there are certain limitations to this analytical approach.[24] The information gathered by chemical extraction reflects the average contents in bulk soils, however it gives no detailed insight into the distribution or the actual status of the particles.

In this study, we took the opportunity to search for, and extract, very small (5 - 100 µm) single particles in the environment of the Chernobyl exclusion zone from three different sampling points around the reactor in 2015 and 2017. Based on sets of individual particles, we showcase the differences in the particulate inventory as it currently stands at these locations. Also, we propose a 'best practice' procedure for the effective extraction of these particles and preparation of single particle samples ready to analyse with destructive (e.g. leaching experiments) as well as modern and non-destructive (e.g. single-particle gamma spectroscopy, SEM, EDS, SIMS) methods.

2 MATERIALS AND METHODS

2.1 Sampling

To compare the effectiveness of particle extraction procedures for different environmental matrices, three sampling points in the vicinity of the former nuclear power plant were chosen (see Figure 1).

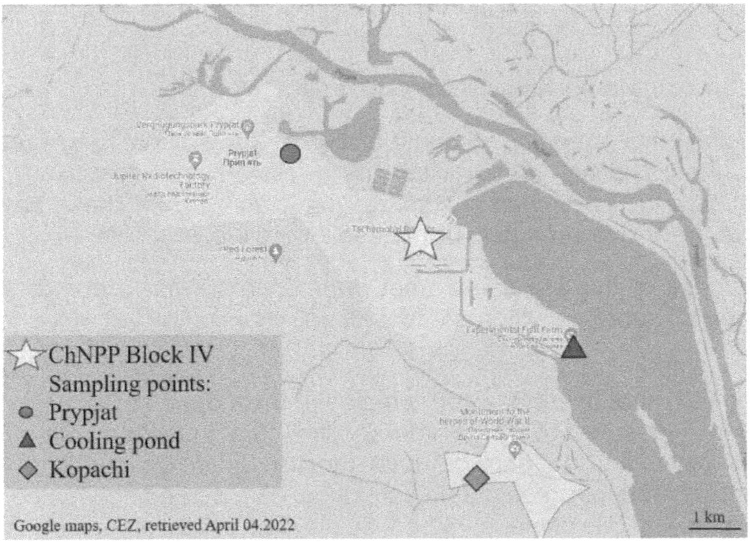

Figure 1 *Locations of sampling sites of soil, sediment and around Chernobyl Nuclear Power Plant. Star = Chernobyl Nuclear Power Plant, Circle = Prypjat asphalt, Diamond = Kopachi soil, Triangle = Cooling pond sediment.*

Particles from Kopachi (51°20'54" N; 30°07'41" E) were isolated from top-soil samples taken from the BioVeStRa (Biological methods for precautionary radiation protection against radionuclide contamination) test field in two sampling campaigns in May and September 2017.[25] The field is situated on the former agricultural land of the local farming community of Kopachi Village, 5 km south-southeast of the ChNPP (Figure 1). The soil can be characterised as a slightly acidic (pH($CaCl_2$) = 5.6), very silty sand, with limited content of clay and organic matter, resulting in a low cation exchange capacity (CEC). For a detailed pedological characterisation.[26] A total number of 140 random surface soil samples (0 - 5 cm depth) of roughly 40 g dry weight each were taken and searched individually for particles.

A Cooling Pond (51°22'28.58"N; 30° 8'34.39"E) sediment sample of roughly 200 g with 15 cm^2 surface area and 10 cm depth was obtained in May 2017 from 10 cm below the water level (pH = 7.5) at the west bank of the cooling pond.

Prypjat samples consisted of two drill cores from the asphalt of the marketplace roughly 3.5 km north-west of the reactor in 2015 (51°24'21.36"N;

30° 3'28.54"E) and had a diameter of 3 cm. The sampled asphalt had been identified as a radiometric hotspot and was thus chosen for further investigation. The pH value of the drill core is mainly influenced by the rain and therefore fluctuates in the slightly acidic range. The drill core was not treated further before searching the particles.

Soil and cooling pond samples were oven dried (105°C) until weight constancy was reached, and then sieved using a 2 mm mesh prior to further analysis.

2.2 Isolating single particles from bulk soil

2.2.1 Radiometric screening. Extensive studies have been conducted in the past on the isolation of particles using radiometric screening for the identification and isolation of radioactive particles.[23,24,27,28] These have been adapted to the above sample sets in the following ways.

In the case of the soil sample set from Kopachi with a comparably low particle density a pre-selection process of soil was conducted in the following way:

Nearly 140 random surface soil samples (0-5 cm depth) of roughly 40 g dry weight were measured using gamma spectroscopy in a petri dish geometry. From this screening, 15 samples with the highest ^{137}Cs count rates were chosen for an in-depth radiometric screening. The screening was conducted using an overlying grid of measurement points. The soil samples were filled into a polyethylene bag, spread to 0.5 cm thickness, and grid lines 1.5 cm x 1.5 cm) were drawn on the polyethylene bag surface. The center of each grid was monitored by using the Geiger counter with a measurement time of 1 minute per spot, see Figure 2.

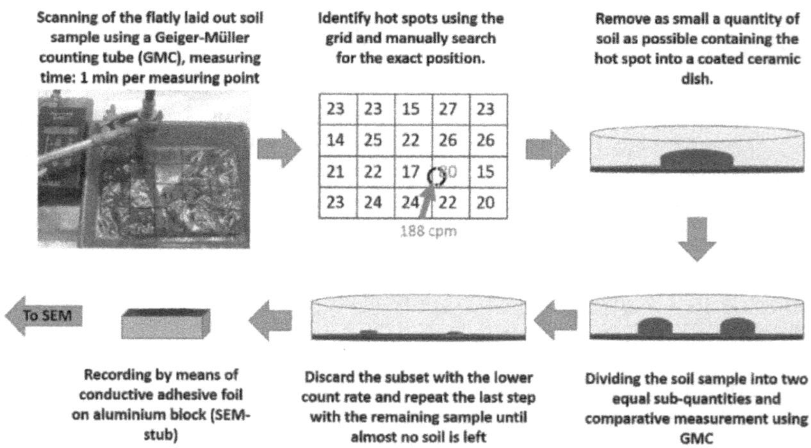

Figure 2 *Scheme of radiometric scanning.*

When the spot of highest activity per minute was located, the underlying soil was separated and the above procedure of binary splitting was applied.

A small amount (roughly 1 - 2 g) of soil was carefully separated and put in a plastic dish. To isolate the fuel particles from any soil matrix, this soil was then searched using binary splitting.[28] The soil was split in two equal halves and separated inside the plastic dish using a spatula. The activities of both sub-samples were measured using a Geiger counter with a 1.5 - 2.0 mg/cm^2 mica window and a diameter of 1.5 cm, and the remaining soil of lower activity was discarded. This process was repeated until practically no soil was visible around the hot spot of activity.

After several iterations of the binary splitting process the hot particle was sufficiently isolated from any soil matrix and the hot spot was carefully extracted using an SEM-stub with double-sided sticky carbon tape. Successful uptake was subsequently confirmed using the Geiger counter and the SEM-sample forwarded for further isolation using a tungsten needle (see. 2.2.3.).

2.2.2 Flotation. An alternative approach for the isolation of hot particles from environmental samples is through flotation. This can be done by using high density liquids such as bromoform,[10,13,29] or through solutions of chemically less hazardous polytungstate solutions.[30,31,32]

Cooling pond sediment and soil samples were incrementally sieved with a mesh size of 63 µm. About 5 g of the sieved sample was loaded into a 15 ml centrifuge tube with 10 ml of a high-density polytungstate solution (δ =3 g/cm^3). To distribute the sample homogeneously in the solution, the centrifuge tubes were vortexed and sonicated in an ultrasonic bath for 5 minutes. To separate the "lightweight" (< 3 g/cm^3) from the heavy fraction, the tubes were centrifuged at 179 g for 30 min. The heavy soil particles – presumably containing all uranium phases (δ(UO_2) = 10.97 g/cm^3) - sedimented at the bottom, whilst the lighter fraction floated on top of the polytungstate solution, see Figure 3.

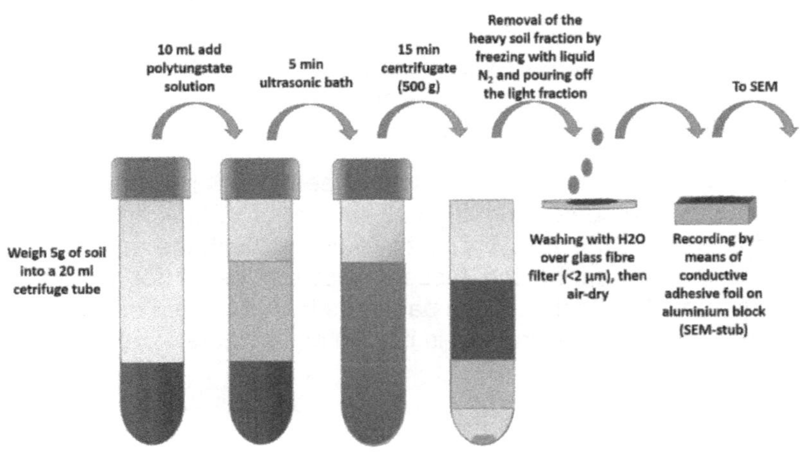

Figure 3 *Scheme of the flotation method.*

The bottom of the tube was then frozen with liquid nitrogen and the supernatant discarded. The remaining particles were filtered using a glass fibre filter (0.5 µm pore size) and washed with water. From the surface of the dried filter, the heavy particles were transferred to a sticky carbon tape on an aluminium sample holder for further screening via SEM.

The flotation method as described here allows less throughput in terms of mass or volume of soil/sediment compared to the radiometric method. Although it necessitates more equipment and time per run than the radiometric screening, only a single run is needed to separate particles. Its major advantage over the radiometric method is its independence from particle or matrix activity and particle size, as the separation solely relies on density.

Due to the need for rinsing of the heavy fraction from adherent solution on a filter, another drawback of the method using polytungstate solutions lies in the inefficiency of the recovery of particles from these filters. Here, the uptake of the particulate fraction using sticky tape is often incomplete as very small particles can be lost in the filter's fibres. Additionally, it should be noted that a comparably large number of non-uranium high-density particles in soils (e.g. containing lead or europium) are co-separated in the process, yielding congested SEM samples for future identification of the uranium particles of interest.

2.2.3 Extraction of single particles on tungsten needles. Using an ESEM (Philips XL30, remX GmbH, Bruchsal, Germany) equipped with an EDS detector (SDD-Detector, remX GmbH, Bruchsal, Germany), the carbon tape area with the extracted material was mapped with the backscattered electron (BSE). In this imaging mode, particles containing heavy elements like uranium were quickly distinguished from other soil or sediment particles. EDS measurements confirmed the uranium content and gave access to the particle's elemental composition.[33]

To allow further process and to remove background signal, individual particles were extracted from the carbon tape using custom-made tungsten needles mounted onto a micromanipulator (Kleindiek Nanotechnik MM3A, Reutlingen, Germany). An electron beam curing glue (SEMglu, Kleindiek Nanotechnik Reutlingen, Germany) was used to attach particles to the needles.

To extract a particle, a tungsten needle is mounted onto the micromanipulator and coated with a small amount of SEMglu inside the SEM. The glue-coated needle is then brought into contact with the hot particle, and subsequently illuminated by the focused electron beam for curing (ca. 5 minutes). The needle with the affixed particle is then retracted from the sample holder.[29] An example of this process is presented in Figure 4.

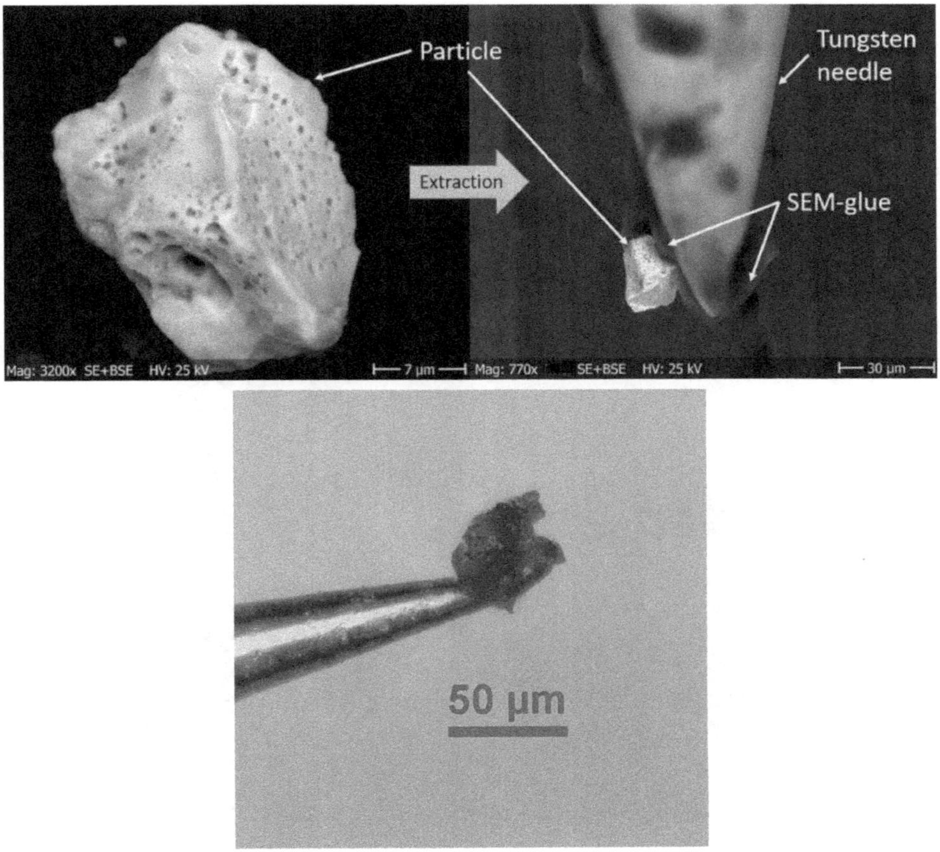

Figure 4 *Particle separation. Extracting the particle from the carbon tape by means of SEMglu and a custom-made tungsten needle, mounted to a micromanipulator. After extraction, the needle with the particle can be removed for further processing. a) The particle on carbon tape before extraction. b) The particle fixed to a custom-made tungsten needle, already mounted onto a sample holder. c) Particle image taken by optical microscopy.*

After extraction, the needle is secured in a custom sample holder with the particle oriented upwards for imaging, EDS mapping and further analysis. High-resolution SEM imaging was performed at 25 kV and smaller spot sizes. EDS measurements were performed at 15 kV and larger spot sizes for improved signal intensity.

The size of the particles was estimated using SEM imaging, measuring the particles two-dimensionally and estimating the third dimension based on the structure. The shape is then approximated to either a sphere or a cuboid to calculate its volume.

2.2.4 Leaching of hot particles. EDS-measurements are qualitative rather than quantitative. To determine the absolute amount of uranium, as well as identify low-abundance elements inaccessible to EDS such as plutonium or fission and activation products, other methods are required. Sequential leaching followed by gamma spectrometry and inductively-coupled mass spectrometry (ICP-MS) was conducted to determine the dissolution process of the particles. A leaching scheme was developed that successively introduces different acids in a step-wise manner, allowing for intermittent observation of the morphological state of the particle via SEM.[34] The first step of the leaching scheme is concentrated HCl, which has non-oxidising properties. It should not dissolve UO_2 but could leach the more soluble elements like ^{137}Cs. Next steps are a diluted and a concentrated HNO_3, which have oxidising effects on the particle. It is expected that the structure of the particle changes.

Figure 5 *Experimental Setup for Leaching of a single particle. The schematic structure of the leaching can be seen in a). The experimental set-up in the laboratory is shown in b). Only the needle with the particle at the tip is held in a Teflon needle holder in the respective acid.*

3 RESULTS AND DISCUSSION

3.1 Optimised procedure for single-particle extraction

In the ^{137}Cs-contaminated soil of Kopachi, a minimum ^{137}Cs activity of 3 - 4 Bq (and mainly 15 µm x 15 µm x 15 µm in size) per particle was required for the method to be successful. However, the chances for isolation of a given particle dropped sharply with lower activity particles and the procedure could only be successfully applied after a pre-concentration. The method could also be very successfully applied to the asphalt sample that contained high-activity particles at the surface in little surrounding matrix. In the case of cooling pond sediment, however, the method was poorly practical. The content of particles per volume was too high, resulting in high overall background, and the ^{137}Cs activity concentration in each particle was relatively low (see results).

While radiometric screening is a fast and simple method for searching sample quantities of any size it is limited to the finding of particles of elevated activity relative to the matrix. The flotation method on the other hand can reliably separate uranium-based high-density particles from soil or sediment matrices. However, it cannot be used to separate particles with a density similar to or lower than the matrices' density. This rules out using flotation to find secondary or mixed uranium-particles like the three outliers discussed in the next subsection.

Based on our experience with the current contamination situation in Chernobyl, we therefore suggest the following procedure as an optimal approach for highest efficiency in terms of reducing sample volumes to SEM-compatibility;

First, the dried sample matrix should be incrementally sieved and – depending on density of particle contamination – pre-treated to identify subsamples of increased activity. The sample should then be radiometrically scanned (see 2.2.1) to extract particles of high relative activity. This is also possible for the sediment samples with high background and lower ^{137}Cs-activity per particle to separate those with unusual high activity.

The subsequent application of the flotation method (see 2.2.2) allows for the finding of high-density particles of any size and activity.

3.2 Identified particles

A full list of identified particles with their given sample codes and sampling location is provided in Table 1.

Table 1 *Distribution of identified particle types found in the three sampling locations/matrices. * Sub-micron nodules composed of iron and sulfur attached to particle. ** Too small for assessment based on SEM picture.*

Particle type	Number of hot particles		
	Sampling location		
	Kopachi	Cooling Pond	Prypjat
UO_2	4	3 (1+2*)	4
$UO_{(2+x)}$	1	6	6
$U_XZr_YO_Z$	1	2	0
Unclassified	3	1*	1**

The majority of particles identified could be ascribed to one of the three main classes of particles suggested by Kashparov et al. and Salbu et al.[10,12] Example images are presented in Figure 4. Particle *Herakles*[1], in Figure 6a, is an example of a UO_2 particle, exhibiting the characteristic porous structure strongly resembling the original fuel. *Ares*, Figure 6b, shows a similarly porous structure on parts of its surface, but with a large area of a smoother coating typical for an annealing with zircaloy. *Johanna*, Figure 6c, does not show the original porous structure in contrast, and rather matches the description of oxidised particles. However, it should be noted that only a very rough estimate on the state of oxidation can be made based solely on the particle's morphology. Additionally, one should consider that the extent of oxidation can be gradual.

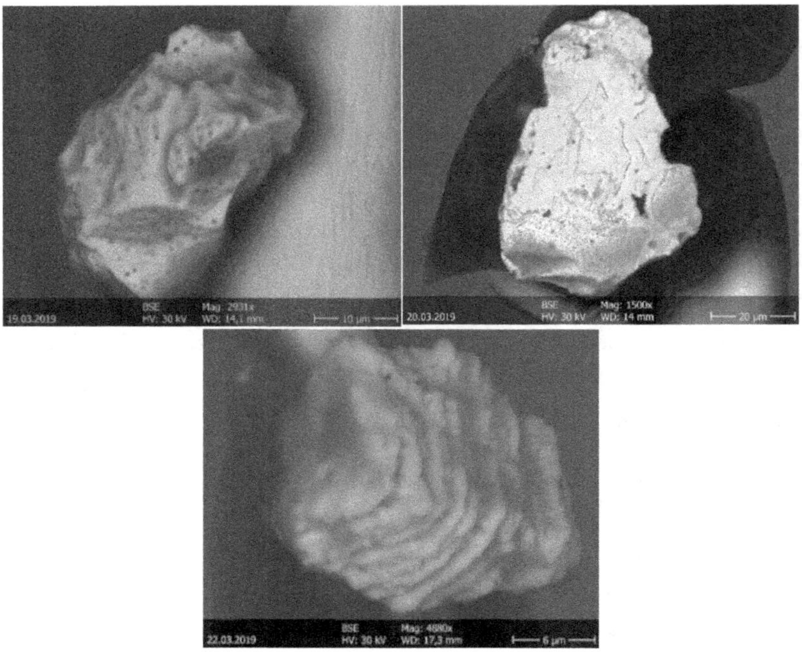

Figure 6 *Particle classes. Example SEM-images of particle classes. The particles are on tungsten needles for better measurement results. The classification into the classes is made purely visually on the basis of the characteristics of these groups described in the literature and additionally EDS measurements, see Section 1. a) Image of Herakles from the Kopachi sampling site with the characteristic pores that are typical for* UO_2 *particles. The EDS can be found in Figure S1. b)* $U_XZr_YO_Z$ *particle Ares from Kopachi, The EDS can be found in Figure S2. c) Oxidized particle (*UO_{2+X}*) Johanna from the Prypjat drill core. The EDS can be found in Figure S3.*

[1] All particles are named based on their respective sampling locations. Particles from Kopachi have names derived from Greek mythology. Names from Norse mythology refer to cooling pond particles while all other names are associated with the Prypjat sampling location.

Although most identified particles could be associated with one of the three main classes of environmental particles (Figure 6), some showed features that did not allow for an allocation into these categories. A number of cooling pond particles had characteristic sub-micron sized nodules rich in iron and sulfide attached to their surface (presumably biogenic FeS_2).[35,36] In one other case, this attachment was so dense that it prevented a clear assessment of the underlying particle type. Additionally, three particles from the Kopachi site showed distinct features that cannot be associated with any of the three main classes.

Based on the three particle sets presented in Table 1, three observations can be made. Firstly, presumably higher oxidised particles ($UO_{(2+x)}$) can still be found in the cooling pond, as well as in the asphalt sample from Prypjat, but are rarely found in the soil sample from Kopachi. This finding supports the reports of higher dissolution rates for oxidised particles in soils.[13] The relative absence of oxygen at the bottom of the cooling pond, as well as the different oxidation conditions of the asphalt samples compared to soil samples may effectively prevent the full dissolution of this particle type (for dissolution rates in the cooling pond sediment, see also Konoplev et al.[37]

Secondly, a number of particles found in the cooling pond's sediment show attachments of micron to sub-micron-sized nodules containing iron and sulfur (Figure 7). Their features closely resemble sedimentary FeS_2 crystals that have been reported to be formed by magnetotactic bacteria under comparable hydrochemical conditions.[35] Their presence is therefore a strong indicator of the anoxic and reducing conditions in the cooling pond sediment zone, responsible for the low dissolution rates of fuel particles in this environment.

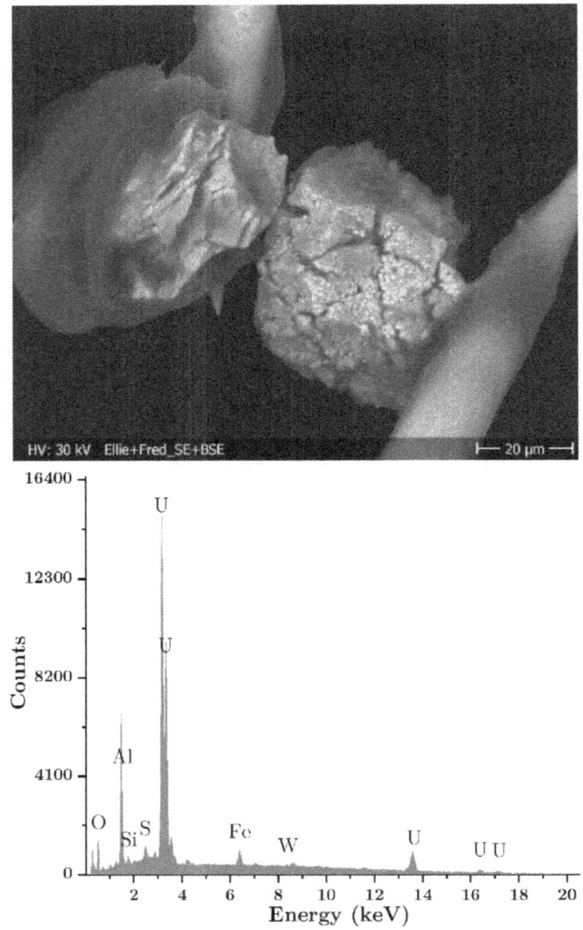

Figure 7 *Particles and EDS-spectrum, a) Two particles from the cooling pond on tungsten needles. A uranium particle with characteristic nodules of iron and sulfur covering its entire surface on the right, in comparison to one without such nodules on the left. In b) is the EDS-spectra for the left particle. The peaks from uranium, tungsten from the needle, sulpha and iron are seen.*

3.3 Leaching results

The three particle types according to Table 1 are expected to differ not only in morphology and composition but also in chemical stability. For direct chemical investigations the particles, mounted on the tungsten needles, were submersed into different liquids in a sequential leaching process. The following leachates were chosen: concentrated HCl, dilute HNO_3 and concentrated HNO_3 in order to understand weathering behaviour and bioavailability of the radionuclide content of the particles. After variable residence time in the liquid, the particle was taken out again and the concentration of ^{137}Cs, ^{154}Eu, ^{241}Am and the Pu isotopes was determined.

Before the procedure, and after each sequential step, SEM pictures of the particles were recorded (Figure 8). After the particle was treated with the three different acids (HCl, HNO_3 at different concentrations), the solutions obtained were analysed using gamma spectroscopy and ICP-MS. There is a recognisable dissolution process of the particle, with morphological changes documented via intermittent SEM images. To ensure that all possible solid parts were dissolved prior to ICP-MS analysis, the final solution was treated with HF.

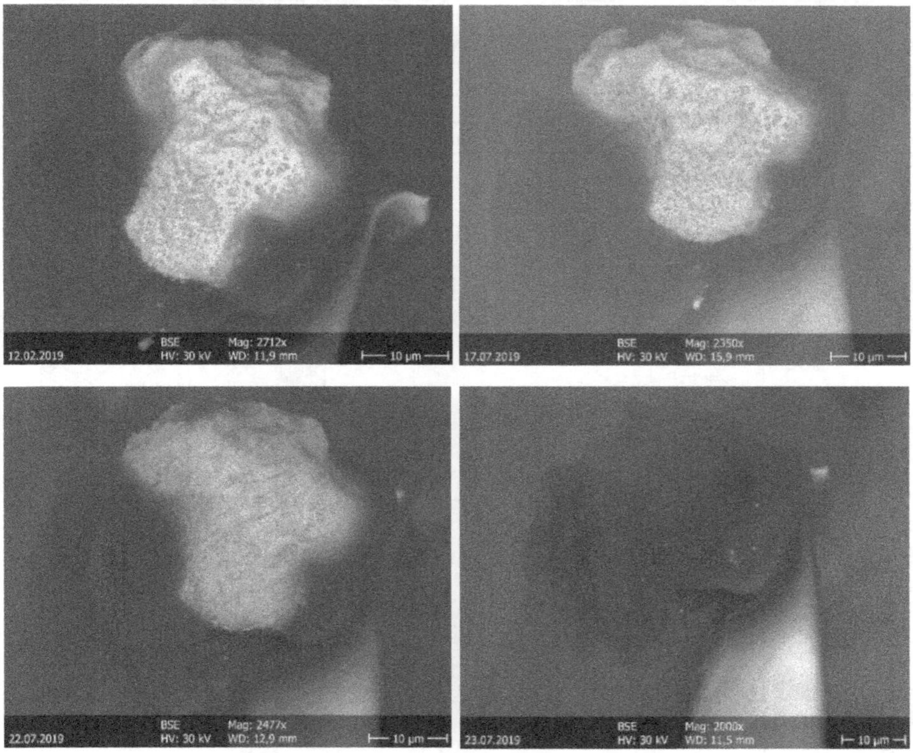

Figure 8 *Leaching Process. This illustration shows the process of dissolution. In picture a, the particle has only been extracted and has not been in further contact with the acids. Picture b shows the particle after leaching with HCl. No structural changes are visible. Picture c shows the particle after a few hours in the diluted HNO_3. Clear dissolution processes are visible. After a few more hours in concentrated HNO_3, the particle is completely dissolved or has fallen off the needle - picture d.*

No visible changes in morphology were observed after the first step in HCl, though through subsequent analysis it was evident that small amounts of ^{137}Cs, ^{241}Am and the Pu isotopes were dissolved (see Table 2 and Figure 9). Significantly more leaching occurs in the second step (diluted HNO_3) as expected. The majority of leaching occurs in the final step (concentrated HNO_3), where the particle either dissolves completely or falls off the needle, and only the SEMglu and needle remain. As was also to be expected here, the highest proportion of fission and activation products was measured in this fraction (see Table 2 and Figure 9). As a proof of concept, this leaching method is shown capable of analysis of the absolute plutonium concentration in individual particles.[34]

Table 2 *Concentrations and masses of the various plutonium isotopes in the different acids. Reproduced from ref. 34 with permission from M. Weiß.*

Nuclide	^{239}Pu			^{240}Pu			^{241}Pu		
	LOD=0.04 ppt			LOD=0.05 ppt			LOD=0.08 ppt		
	ppt	pg	±	ppt	pg	±	ppt	pg	±
Blank	0.25	2.68	37.40%	0.2	2.13	38.70%	<LOD	0.66	91.80%
HCl	0.8	8.27	27.10%	0.47	4.86	28.00%	<LOD	0.71	72.20%
HNO_3 dil.	6.5	68.64	9.40%	2.96	31.25	11.20%	0.22	2.35	47.50%
HNO_3 conc.	10.57	115.72	7.00%	4.86	53.28	10.30%	0.33	3.6	40.50%
Total	17.1	184.61	8.60%	7.68	82.99	11.60%	0.43	4.68	53.50%

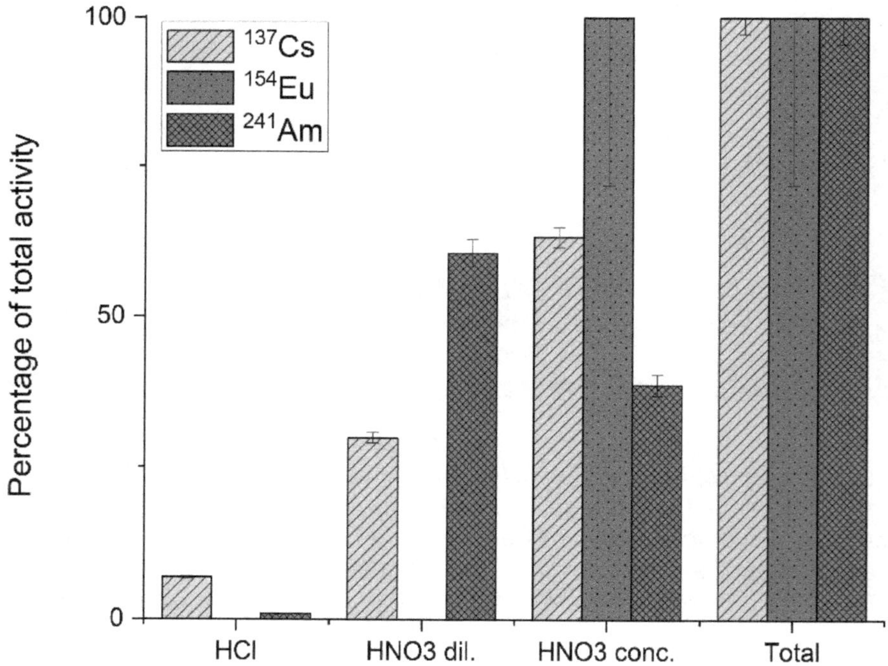

Figure 9 *Activity of the leaching solutions measured by gamma counting. The activity of ^{154}Eu was below the detection limit in the first two solutions. ^{137}Cs was dissolved out in all three acids as expected since it is an easily soluble element. ^{241}Am was dissolved mainly in the oxidizing acid, where the particle also changed structurally.*

4 CONCLUSIONS

This work confirms the ongoing contamination of the Chernobyl area due to weathering of small hot particles with very specific individual characteristics dependent on the various environmental matrices and directions relative to the reactor.

The proposed process of searching for particles in different matrices and the fixation of particles to needle tips allows for uncomplicated handling and paves the way to in-depth analysis of particles using various methods. Particle classification based solely on SEM imaging matches the classification proposed by Kashparov et al. and Salbu et al. in most cases,[10,12] however, especially in the case of Kopachi-soil samples, a number of particles was found that could not be ascribed to any of these known particle classes, giving rise to future scientific investigations. The proof of concept for leaching experiments provided in this work means that further weathering analyses and leaching experiments on individual particles can follow.

5 OUTLOOK

After the particles have been separated, various analyses are now carried out.

Whilst this work concentrated on the extraction and compositional comparison between the sampling sites, further projects will also investigate individual isotope signatures and further radioecologically relevant elements like the actinides and ^{90}Sr.

Due to the wide variation of chemical behaviour, further investigation into the different particle types and the special outliers is needed. Future projects should investigate sequential individual particle leaching behaviour to contribute detailed data to established radionuclide mobility models.

Overall, the sample set presented in this work has to be significantly widened in order to answer the question of whether or not the presented secondary particle types are of considerable significance to the present and future contamination situation in the Chernobyl Exclusion Zone.

6 ACKNOWLEDGEMENTS

Financial support of this work from the German Federal Ministry of Education and Research under the contract numbers 02S9276D and 02NUK044A is gratefully acknowledged. GSt and SD gratefully acknowledge partial financial support by the Prof. Joachim Lenz Foundation. We also thank our interns Leo Wujec and Jan Karcewsky for their great help in searching and isolating particles from dozens of soil samples. We also thank Darcy van Eerten for grammar and spellchecking of the manuscript.

7 REFERENCES

1. Y. Satou et al. *Anthropocene,* 2016, **14**, 71.

2. G. Steinhauser, *Journal of Radioanalytical and Nuclear Chemistry,* 2018, **318**, 1629.
3. Y. Igarashi et al. *Journal of Environmental Radioactivity,* 2019, **205**, 101.
4. O. Lind et al. *Science of the Total Environment.* 2006, **376**, 294.
5. V. Kashparov et al. *Science of the Total Environment.* 2003, **317**, 105.
6. G. Steinhauser, A. Brandl and T. Johnson, *Science of the Total environment.* 2014, **470-471**, 800.
7. UNSCEAR. 2010. *Report of the United Nations Scientific Committee on the Effects of Atomic Radiation 2010.* s.l.: UN Publications, 2010. Vol. M.11.IX.4.
8. E. Grishanin, *Physics of Atomic Nuclei.* 2010, **73**, 2296.
9. V. Kashparov, *Nuclear Technology.* 1996, **2**, 246.
10. V. Kashparov, *Applied Geochemistry,* 2012, **27***,* 1348.
11. C. Walther and M. Denecke, *Chemical Reviews,* 2013, **113**, 995.
12. B. Salbu et al., *Journal of Environmental Radioactivity,* 2018, **186**, 101.
13. V. Kashparov et al., *Journal of Environmental Radioactivity.* 2004, **72**, 335.
14. V. Kashparov, S. Levchuk et al., *Earth System Science Data,* 2018, **10**, 339.
15. V. Kashparov et al., *Journal of Nuclear Materials.* 2000, **279**, 225.
16. P. Schubert and U. Behrend, *Radiochimica Acta.* 1987, **41,** 149.
17. R. Pöllanen, I. Valkama and H. Toivonen, *Atmospheric Environment.* 1997, **31,** 3575.
18. L. Devell et al., *Nature,* 1986, **321***,* 192.
19. V. Kashparov et al., *Journal of Environmental Radioactivity,* 2020, **106387**, 223.
20. W. Runde, M. Neu and G. Goff, *Encyclopedia of Inorganic Chemistry.* 2010, **19**, 4.
21. V. Kashparov et al., *Health Physics,* 1999, **76**, 251.
22. V. Kashparov et al., *Journal of Environmental Radioactivity,* 2019, **208-209**, 106025.
23. IAEA. Radioactive Particles in the Environment: Sources, Particle Characterization and Analytical Techniques. *IAEA-TECDOC-1663.* 2011.
24. R. Pöllanen et al., *Journal of Environmental Radioactivity,* 1999, **45**, 149.
25. L. Traxler, *Journal of Hazardous Materials.* 2021, **403**, 124002.
26. W. Schulz et al., *Applied Geochemistry,* 2019, **101**, 103.
27. B. Salbu, T. Krekling and D. Oughton, *Analyst,* 1998, **123**, 843.
28. P. Burns et al., *Applied Radiation and Isotopes,* 1995*,* **46**, 1099.
29. U. Admon. *Radioactive particles in the Environment, Single Particles Handling and Analyses.* [ed.] V. Kashparov, D. Oughton. The Netherlands: Springer, 2007.
30. S. Crow et al., *Biogeochemistry,* 2007, **85**, 69.
31. X. Zou et al., *Environmental Geology.* 2006, **5**, 813.
32. B. Plewinsky and R. Kamps, Sodium metatungstate, a new medium for binary and ternary density gradient centrifugation. USGS Report1984, 1429j1439.

33. O. Lind et al., Overview of sources of radioactive particles of Nordic relevance as well as a short description of available particle characterisation techniques. *NKS-180.* 2008.
34. M. Weiß, *PhD Thesis.* Hannover: s.n., 2022.
35. J. Schieber, Iron Sulfide Formation. [book auth.] J Reitner and V Thiel. *Encyclopedia of Geobiology. Encyclopedia of Earth Sciences Series.* Dordrecht: Springer, 2011.
36. M. Pósfai et al. *American Mineralogist,* 1998, 83, 1469.
37. A. Konoplev et al., *The Analyst.* 1992, 117, 1041.

8 SUPPLEMENTARY FIGURES

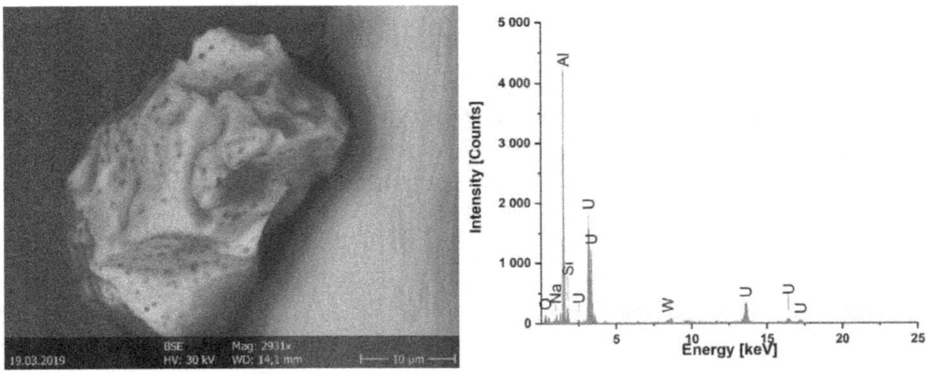

Figure S1 *a) Heraklion is a classic* UO_2 *particle with pores. Please see ref. 12 for a comparison of images with the literature. The similarities and EDS measurements suggest a comparable* UO_2 *particle. b) EDS von Heraklion with uranium and tungsten from the Needle.*

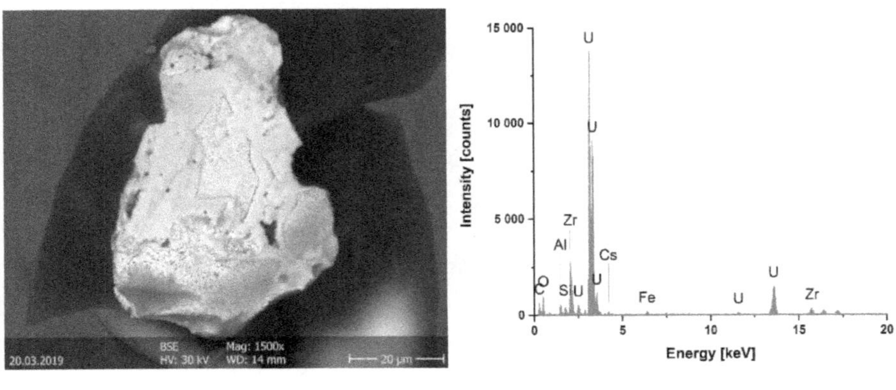

Figure S2 *a) Ares is a* $U_XZr_YO_Z$ *particle. Please see ref. 12 for a comparison of images with the literature. The similarities and EDS measurements suggest a comparable Zr particle. b) EDS from Ares with Zr-Peaks.*

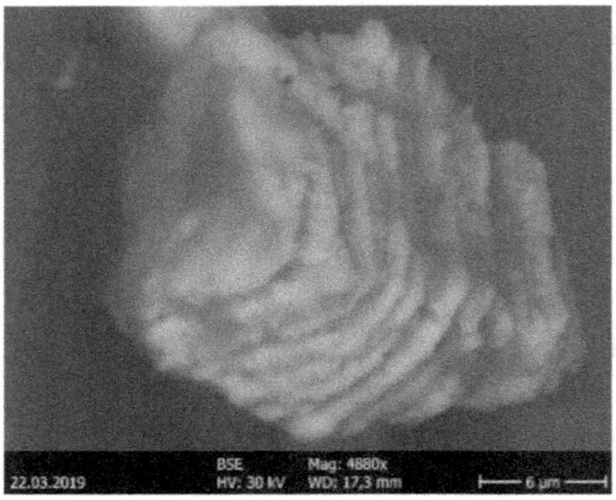

a)

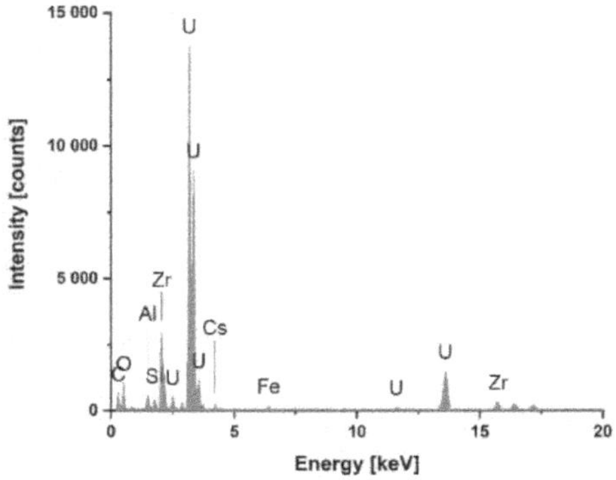

b)

Figure S3 *a) Johanna is a particle of the UO_{2+x} particles. Please see ref. 12 for a comparison of images with the literature. The similarities and EDS measurements suggest a comparable UO_{2+x} particle. b) EDS from Johanna with mainly uranium and some parts of aluminium and silicon.*

CHAPTER 2 DETERMINATION OF TRANSURANIUM ELEMENTS PRODUCED BY THE CASTLE BRAVO EXPLOSION

J. A. Corcho-Alvarado[a*], S. Röllin[a] and H. Sahli[a]

[a]Nuclear Chemistry Division, Spiez Laboratory, FOCP, Switzerland

*jose.corcho@babs.admin.ch

1 INTRODUCTION

In a thermonuclear explosion, a target (e.g. uranium) is exposed to an intense neutron flux with the resultant multiple neutron capture reactions and subsequent beta decays, producing transuranium elements.[1-3] Hence, thermonuclear explosions offered a possibility for investigating the production of new isotopes of known, or new, transuranium elements.[2-8] The thermonuclear tests "Mike", fired at Enewetak (Marshall Islands) and "Barbel" and "Par", at Nevada test site (USA), were among the most investigated ones.[2,5,6,7,9,10] The thermonuclear explosion "Mike", in November 1952, was the first test where the production of transuranium elements was investigated.[2,4,9] Debris from this explosion was found to be rich in transuranium elements up to ^{255}Fm.[3,4] These studies demonstrated, for the first time, the presence of the isotopes ^{244}Pu, ^{246}Pu and ^{246}Am.[3,4] The occurrence of these isotopes suggested that very heavy uranium isotopes (e.g. ^{239}U, ^{240}U...^{255}U) were formed in the explosion. These heavy uranium isotopes then decayed by beta emission to more stable isobars. The production yield of the transuranium nuclides was shown to decline with increasing mass according to an exponential trend. A slightly higher abundance of the nuclides with even mass until mass 250 was also observed. Even mass abundances were shown to be slightly higher than odd-mass abundances by a factor of 1.58 ± 0.14.[2] In another study of the "Mike" explosion, elements 94 through 100 were chemically extracted and purified from airborne and condensed debris samples.[2] The even-odd effect was explained by differences in the neutron capture cross section.[2] The process of nuclei formation in thermonuclear explosion has been also modelled using kinetic models describing stellar nucleosynthesis.[11,12] In these models, the time-dependent neutron fluxes have been reduced to values < 10^{-6} s and the processes accompanying the β^- decay of neutron-rich nuclei has been included. The even-odd anomaly in the observed yields of production of the nuclei has been explained by the prevailing effect of processes accompanying the beta decay of heavy isotopes.[12]

While published studies have covered numerous thermonuclear explosions, little information has been reported for important tests such as the "Castle Bravo" explosion. This thermonuclear test was fired by USA on 1 March 1954 at Bikini Atoll (Marshall Islands) and produced the largest yield (15 MT) of the entire series in the Pacific Proving Grounds (PPG), about 1000 times more powerful than the atomic bomb dropped on Hiroshima in 1945.[13] The device for Castle Bravo had the same basic configuration as the Mike device but with a different type of fusion fuel. Whereas the Mike device used cryogenic liquid deuterium, the Castle Bravo device used lithium deuteride (LiH). The device designers, in a prediction error, considered only the isotope ^{6}Li as reactive. The isotope ^{7}Li, which accounts for about 60% of the Li content in the LiH, was considered inactive to neutron reactions. However, now is known that when ^{7}Li is bombarded with energetic neutrons, the nucleus instantly decays into an alpha particle, a ^{3}H nucleus and another neutron. As a consequence of this reaction, much more ^{3}H was produced than expected, with extra ^{3}H fusing with ^{2}H to produce extra neutrons ($^2H + {}^3H \rightarrow {}^4He + n$) of monoenergetic energy (E_n = 14.1 MeV). The occurrence of these reactions then resulted in a much larger neutron flux. The result was a greatly increased fissioning of the U tamper and thus, a much higher- than-expected yield (about 2.5) for the explosion.[14] The high yield along with a powerful eastward wind current, led to the most significant accidental radioactive contamination ever caused by the US.[15] Large quantities of radioactive debris were released into the atmosphere, and spread by the prevailing winds over a large area. Fallout from the explosion fell on residents of Rongelap and Utirik atolls some 120 km away and quickly spread around the world. The cloud contaminated more than 18 000 km^2 of the surrounding Pacific Ocean, including the surrounding atolls of Rongerik, Rongelap and Utirik.[15]

For the "Castle Bravo" test, a U tamper was employed. The high-energy neutrons produced by the fusion reaction caused the ^{238}U to undergo both fission and neutron capture, generating ^{239}Pu as:

$$^{238}_{92}\mathrm{U}(\mathrm{n},\gamma)^{239}_{92}\mathrm{U}(\beta^-)\xrightarrow{23.4\ \mathrm{m}}{}^{239}_{93}\mathrm{Np}(\beta^-)\xrightarrow{2.4\ \mathrm{d}}{}^{239}_{94}\mathrm{Pu}$$

The heavier Pu isotopes were then produced mainly via successive neutron capture reactions on uranium isotopes and subsequent β-decay to the corresponding Pu isobars. Most of the Pu isotopes, with the exception of ^{241}Pu (half-life: 14.33 y) and ^{243}Pu (half-life of 4.956 h), are characterized by a long half-life: ^{239}Pu (half-life of 24.1 x 10^{3} y), ^{240}Pu (half-life of 6.56 x 10^{3} y), ^{242}Pu (half-life of 3.73 x 10^{5} y) and ^{244}Pu (8.11 x 10^{7} y). Following β-decay of ^{241}Pu and ^{243}Pu, ^{241}Am (half-life of 432.6 y) and ^{243}Am (half-life of 7.367 x 10^{3} y), respectively, are formed. The amounts and isotopic compositions of U and transuranium elements left after the explosion vary according to the completeness of the fission chain. Apart from the analysis of the so-called "Bikini ash" debris and marine sediment samples collected at Bikini,[14,16,17,18,19] little has been published about the production of transuranium isotopes in this thermonuclear explosion. This may be related to the fact that the Bravo bomb vaporized an artificial island, leaving a 75 m-deep and 1.5 km-diameter crater, where six more nuclear weapons were detonated.[14] Hence, radioactive contamination in the Bravo crater stems from a mix of nuclear weapons.

We aimed at determining the yield of long-lived transuranium radionuclides (e.g. ^{239}Pu, ^{240}Pu, ^{241}Pu, ^{242}Pu, ^{244}Pu, ^{241}Am, ^{243}Am and ^{245}Cm) produced in the "Castle Bravo" thermonuclear explosion. We based our study on the analysis of surface soil samples collected at Bikar, Marshall Islands, an atoll that was contaminated primarily with radioactive fallout debris from the "Castle Bravo" explosion.[20,21] The results of the radionuclide activity concentrations and ratios are discussed in view of the production process. These results are valuable for sample origin and contamination date determination.

2 EXPERIMENTAL METHODS

2.1 Sampling and radiochemical analysis

The Bikar Atoll is located in the far northeast of the Ratak Chain, about 500 km east from Bikini Atoll and 850 km east from Enewetak Atoll, both proving grounds for nuclear weapon testing in the Marshall Islands. Due to its geographical location, Bikar constitutes an important reference site to evaluate the extension of the regional fallout from the testing sites in Bikini and Enewetak. Detailed information about the collection of the soil samples in Bikar Atoll was published elsewhere.[20,21] In October 2019, six surface soil samples (top 10 cm) were collected at Bikar and Almani Islands, Bikar Atoll. One soil profile was collected at Jabwelo Island, at increments of 10 cm up to a total depth of 40 cm. The soil samples were oven dried, screened through a 2 mm stainless steel sieve, ground and well-mixed to produce homogeneous samples. Aliquots of the soil samples were ashed at 520 °C for 24 h. The Pu, Am and Cm radionuclides were analysed as reported elsewhere.[20,21,22,23] Briefly, up to 5 g aliquots of ashed soil were used for the analysis. For determining the Pu, Am and Cm radionuclide activities, known amounts of ^{242}Pu, ^{115}In and ^{243}Am tracers were added as radiochemical recovery tracers. The overall radiochemical yields of Pu and Am varied from 70 to 100%. For the analysis of the minor ^{242}Pu, ^{244}Pu, ^{243}Am and ^{245}Cm, soil aliquots were analysed without spike. For these aliquots, the atom ratios ^{242}Pu/^{239}Pu, ^{244}Pu/^{239}Pu, ^{243}Am/^{241}Am and ^{245}Cm/^{241}Am were measured. The ^{242}Pu, ^{244}Pu, ^{243}Am and ^{245}Cm activity concentrations were calculated then from these ratios, and the individual ^{239}Pu and ^{241}Am activity concentrations. The mass spectrometry (MS) analysis was carried out using a double focusing magnetic sector field inductively coupled plasma mass spectrometer (SF-ICP-MS) Element XR (Thermo Fischer Scientific) as described elsewhere.[22,23]

3 RESULTS AND DISCUSSION

3.1 Isotopic contents of Pu and Am

The activity concentrations of ^{239}Pu, ^{240}Pu, ^{241}Pu, ^{241}Am, ^{242}Pu, ^{243}Am, ^{244}Pu and ^{245}Cm in soils collected at Bikar Atoll are given in Table 1. The activity concentrations of the lighter ^{239}Pu, ^{240}Pu, ^{241}Pu and ^{241}Am radionuclides were found to be in the order of a few Bq kg^{-1}. The heavier ^{242}Pu, ^{243}Am and ^{245}Cm radionuclides showed much lower activity concentrations in the order of a few mBq kg^{-1}. ^{244}Pu was also detected at ultra-trace levels, in the range of a few µBq kg^{-1}. Activity of ^{245}Cm was below the limit of detection in most of the samples. Due to the typically low activity levels in environmental samples, the analysis of ^{242}Pu, ^{243}Am, ^{244}Pu and ^{245}Cm by radiometric techniques is not feasible and mass spectrometry techniques are required. Therefore, these radionuclides are rarely reported in environmental samples.

Pu and Am radionuclides showed similar vertical patterns in soil, with highest activities at the surface layer (profile, Table 1). The similarities suggest that these radionuclides have not suffered significant fractionation during the interval between the time they were deposited on the soil surface and the time the samples were collected. We assume that the relative yield is not affected by chemical fractionation in the soil.

Table 1 *Radionuclide activity concentrations measured in soil samples (fraction < 2 mm) collected at Bikar Atoll. The expanded uncertainties with k=2 are reported. Reference date for decay correction is 19th October 2019.*

Sample	Depth (cm)	Radionuclide activity concentrations, in Bq kg^{-1} dry weight							
		^{239}Pu	^{240}Pu	^{241}Pu	^{242}Pu (x 10^{-3})	^{244}Pu (x 10^{-7})	^{241}Am	^{243}Am (x 10^{-3})	^{245}Cm (x 10^{-3})
Bik-1	0 - 10	3.3 ± 0.1	3.1 ± 0.1	6.5 ± 0.8	2.0 ± 0.1	< 2	4.4 ± 0.2	8.8 ± 2	-
Bik-2	0 - 10	16.8 ± 0.7	15.7 ± 0.6	31 ± 1	10.0 ± 0.5	11.4 ± 2.0	21.6 ± 0.8	41 ± 4	-
Bik-3	0 - 10	14.3 ± 0.5	13.5 ± 0.5	26 ± 1	8.7 ± 0.4	10.1 ± 0.7	18.6 ± 0.7	38 ± 4	-
Bik-4	0 - 10	1.9 ± 0.1	1.8 ± 0.1	3.9 ± 0.3	-	2.2 ± 0.6	3.1 ± 0.1	6.7 ± 2	-
Bik-5	0 - 10	3.9 ± 0.1	3.7 ± 0.1	7.1 ± 0.7	-	2.6 ± 1.5	5.9 ± 0.2	-	-
Bik-6	0 - 10	1.8 ± 0.1	1.7 ± 0.1	3.4 ± 0.3	-	< 2	2.6 ± 0.1	-	-
Profile	0 - 10	35.0 ± 1.7	33.7 ± 1.6	66 ± 3	22 ± 1	26.8 ± 2	47 ± 2	75 ± 7	5 ± 11
	10 - 20	24.1 ± 0.9	23.7 ± 0.9	46 ± 2	15.5 ± 0.8	20.5 ± 3	33 ± 1	60 ± 6	1.3 ± 0.7
	20 - 30	1.35 ± 0.04	1.34 ± 0.04	2.9 ± 0.3	0.82 ± 0.03	1.1 ± 0.9	1.7 ± 0.2	-	-
	30 - 40	0.35 ± 0.01	0.34 ± 0.02	1.0 ± 0.3	-	< 0.9	0.5 ± 0.1	-	-

3.2 Pu and Am isotope ratios

The isotope ratios, normalized to mass 239 (^{239}Pu), are reported in Table 2. The mean ^{240}Pu/^{239}Pu, $^{241/239}$Pu, ^{242}Pu/^{239}Pu and ^{244}Pu/^{239}Pu atom ratios are equal to 0.261 ± 0.005, (1.2 ± 0.2) x 10^{-3}, (9.5 ± 0.5) x 10^{-3} and (2.8 ± 0.6) x 10^{-4}, respectively. The low variability shown by the Pu atom ratios confirms a similar source of contamination (Castle Bravo explosion). ^{241}Am is produced by the decay of ^{241}Pu, which has a relative short life of 14.3 yrs. The amount of ^{241}Am will depend on the initial amount of ^{241}Pu and the time elapsed since release. Assuming that little or no fractionation is occurring after deposition, the ratio ^{241}Am/^{241}Pu has been successfully used for dating purposes. The contamination date based on the ratio ^{241}Am/^{241}Pu was already published elsewhere.[20,21]

^{243}Am/^{241}Am atom ratios varying between 2.8 x 10^{-2} and 3.7 x 10^{-2}, with a mean value of (3.3 ± 0.3) x 10^{-2}, were measured in soil from Bikar (decay corrected for 19.10.2019) (Table 2). A comparison of ^{243}Am/^{241}Am ratios with previously reported ratios is not straightforward. ^{243}Am/^{241}Am ratios measured in this study are, nonetheless, only slightly lower than those found in soil and lichen samples collected near Chernobyl (4.4 to 4.6 x 10^{-2}), about 10 years after the accident.[25] In another study in a soil core collected near Chernobyl, a ^{243}Am/^{241}Am activity ratio of (3.4 ± 0.3) x 10^{-2} was reported (Date for decay correction: 26.04.1998).[26] The measured ^{245}Cm concentrations were very close to the detection limit and can only be used as a rough estimation.

Table 2 *Isotopic abundances of Pu and Am radionuclides in soil contaminated with debris from the "Castle Bravo" explosion, normalized at mass 239. The ^{243}Am/^{241}Am ratio is also reported. The expanded uncertainties with k = 2 are reported. Reference date for decay correction is 19th October 2019.*

Soil sample	Atom ratios								
	^{240}Pu/^{239}Pu	^{241}Pu/^{239}Pu (x 10^{-3})	^{242}Pu/^{239}Pu (x 10^{-3})	^{244}Pu/^{239}Pu (x 10^{-4})	^{241}Am/^{239}Pu (x 10^{-2})	^{243}Am/^{239}Pu (x 10^{-4})	^{243}Am/^{241}Am (x 10^{-2})	^{245}Cm/^{239}Pu (x 10^{-5})	^{245}Cm/^{241}Am (x 10^{-4})
Bik-1	0.259 ± 0.005	1.2 ± 0.2	9.5 ± 0.6	3.3 ± 1.2	2.4 ± 0.1	8.3 ± 0.2	3.4 ± 0.7		-
Bik-2	0.256 ± 0.006	1.10 ± 0.03	9.7 ± 0.3	2.1 ± 0.9	2.3 ± 0.1	7.7 ± 0.6	3.3 ± 0.2	2.6 ± 2.9	*11 ± 12*
Bik-3	0.257 ± 0.008	1.09 ± 0.04	9.6 ± 0.3	2.3 ± 0.5	2.3 ± 0.1	8.2 ± 0.7	3.5 ± 0.3		-
Bik-4	0.259 ± 0.016	1.3 ± 0.1	-	4.1 ± 1.2	3.0 ± 0.2	11 ± 4	3.7 ± 1.2		-
Bik-5	0.260 ± 0.006	1.1 ± 0.1	-	2.2 ± 1.0	-	-	-		
Bik-6	0.257 ± 0.005	1.14 ± 0.08	-	< 4	-	-	-		
Profile	0.263 ± 0.006	1.13 ± 0.03	9.8 ± 0.2	2.7 ± 0.2	2.4 ± 0.1	6.7 ± 0.5	2.8 ± 0.1	*5 ± 11*	*22 ± 46*
Profile	0.269 ± 0.006	1.14 ± 0.05	10.1 ± 0.3	2.9 ± 0.4	2.5 ± 0.1	7.8 ± 0.7	3.1 ± 0.2	1.9 ± 1.0	8 ± 4
Profile	0.270 ± 0.007	1.3 ± 0.1	8.5 ± 1.5	3.6 ± 1.8	2.3 ± 0.2	-	-		
Profile	0.260 ± 0.015	1.6 ± 0.6	-	< 8	2.4 ± 0.5	-	-		
Mean	0.261 ± 0.005	1.2 ± 0.2	9.5 ± 0.5	2.8 ± 0.6	2.5 ± 0.2	8.3 ± 1.5	3.3 ± 0.3	3.3 ± 1.8	14 ± 7

Mean ^{241}Am/^{239}Pu and ^{243}Am/^{239}Pu atom ratios of (2.5 ± 0.3) x 10^{-2} and (8.3 ± 1.5) x 10^{-4}, respectively, were found in soil from Bikar (Table 2). ^{243}Am and

^{239}Pu have relatively long half-lives, and do not seem to suffer significant fractionation in soil from Bikar. Therefore, $^{243}Am/^{239}Pu$ ratios are likely linked to the source of contamination. Data on ^{243}Am in environmental samples is unfortunately very scarce. Only a few studies on ^{243}Am have been published so far. In materials released as a consequence of the Chernobyl accident, for example, a typical $^{243}Am/^{239}Pu$ isotope ratio of 1.8×10^{-3} was reported [26]. In a hot particle found in Pripyat, after the Chernobyl NPP accident, a $^{243}Am/^{239}Pu$ atom ratio of 1.9×10^{-3} was reported.[27] $^{243}Am/^{239}Pu$ isotope ratios in particles released after the Chernobyl accident are about two-fold higher than those found in soil from Bikar.

The mean normalized isotope ratios in soil from Bikar are reported in Table 3 and plotted as function of the mass in Figure 1. For comparison, the isotope ratios for the Mike[2] and the Anacostia[8] nuclear tests, and the ones measured in an Irish Sea sediment (IAEA-135), are also shown. Overall, we observe a rapid decrease of the normalized isotope ratios with the mass of the isotope (Figure 1, Table 3), as observed in other studies.[2,5,8,24] The relative abundances are described by two smooth curves, one for the even-massed abundances that lies uniformly above that for the odd-massed abundances (Figure 1). This is the so called odd-even effect. This effect can even be seen for the sediment Irish Sea Sediment IAEA-135, with the exception that ^{244}Pu is missing, as the Pu originated from nuclear reactors.

Table 3 *Isotopic abundances of Pu and Am radionuclides in soil contaminated with debris from the "Castle Bravo" explosion, normalized at mass 239 (^{239}Pu). The expanded uncertainties with k=2 are reported.*

Nuclide	Atom ratio ($^{A}X/^{239}Pu$)			
	Castle Bravo (This study)	Mike [2]	Anacostia [8]	IAEA-135
^{239}Pu	1.000	1.000 ± 0.004	1.0 ± 0.3	1.000
^{240}Pu	0.261 ± 0.005	0.363 ± 0.004	1.09 ± 0.22	0.207 ± 0.01
^{241}Pu	$(1.2 \pm 0.2) \times 10^{-3}$	$(39 \pm 5) \times 10^{-3}$	-	$(16.7 \pm 0.2) \times 10^{-3}$
^{241}Am	$(2.5 \pm 0.3) \times 10^{-2}$	-	-	$(4.4 \pm 0.2) \times 10^{-2}$
^{242}Pu	$(9.5 \pm 0.5) \times 10^{-3}$	$(19 \pm 3) \times 10^{-3}$	0.14 ± 0.02	$(6.9 \pm 0.2) \times 10^{-3}$
^{243}Am	$(8.3 \pm 1.5) \times 10^{-4}$	$(2.1 \pm 0.4) \times 10^{-3}$	$(22 \pm 1) \times 10^{-3}$	$(1.4 \pm 0.4) \times 10^{-4}$
^{244}Pu	$(2.8 \pm 0.6) \times 10^{-4}$	$(1.18 \pm 0.7) \times 10^{-3}$	$(7 \pm 4) \times 10^{-3}$	$< 5 \times 10^{-5}$
^{245}Cm	$(3.3 \pm 1.8) \times 10^{-5}$	$(1.24 \pm 0.01) \times 10^{-4}$	$(3.0 \pm 0.4) \times 10^{-4}$	-

According to Steier et al.,[24] it is expected that, at time zero after the explosion, the ratio between mass 241 and 239 will be the same as that between mass 242 and 240 as:

$$\left(\frac{^{241}_{94}Pu}{^{239}_{94}Pu}\right)_0 = \left(\frac{^{242}_{94}Pu}{^{240}_{94}Pu}\right) \quad (1)$$

Assuming an exponential decrease for the even and odd mass abundances with the same slope, the isotope ratio $^{241}Pu/^{239}Pu$ can be estimated from the regression curve in Figure 1 as follows.

$$\left(\frac{^{241}_{94}Pu}{^{239}_{94}Pu}\right)_0 = e^{-2\cdot 1.71} \quad (2)$$

With a half-life of 14.35 years, the ^{241}Pu concentration will decrease rather quickly whereas the ^{239}Pu, ^{240}Pu and ^{242}Pu concentrations can be considered constant due to the relative long half-lives. With the isotope ratio $\left(\frac{^{241}_{94}Pu}{^{239}_{94}Pu}\right)_t$ from Table 2, the time T since the thermonuclear explosion can be calculated as follows:

$$T = \frac{-T_{1/2}}{ln2} \cdot ln\left(\frac{\left(\frac{^{241}_{94}Pu}{^{239}_{94}Pu}\right)_0}{\left(\frac{^{241}_{94}Pu}{^{239}_{94}Pu}\right)_t}\right) = \frac{-T_{1/2}}{ln2} \cdot ln\left(\frac{e^{-2\cdot 1.71}}{\left(\frac{^{241}_{94}Pu}{^{239}_{94}Pu}\right)_t}\right) \quad (3)$$

Based on equation (3) the time since the contamination date of the Bikar Atoll was estimated to 69 (± 5) years, which corresponds to a contamination date of 1950.

According to the data available for the Mike Thermonuclear Test,[2,12] the Pu isotope ratios are related by a correlation factor of 0.7451 as:

$$\left(\frac{^{241}_{94}Pu}{^{239}_{94}Pu}\right)_0 = 0.7451\left(\frac{^{242}_{94}Pu}{^{240}_{94}Pu}\right) \quad (4)$$

Using this approach, the time T since the thermonuclear explosion can be calculated as follows:

$$T = \frac{-T_{1/2}}{ln2} \cdot ln\left(0.7451\frac{\frac{^{242}_{94}Pu}{^{240}_{94}Pu}}{\frac{^{241}_{94}Pu}{^{239}_{94}Pu}}\right) \quad (5)$$

Based on equation (5), an age of 65 (± 3) years was estimated, which correspond to a contamination date of 1954. This value agrees remarkably well with the date of the Castle Bravo explosion, known to have been taken place on March 1954.

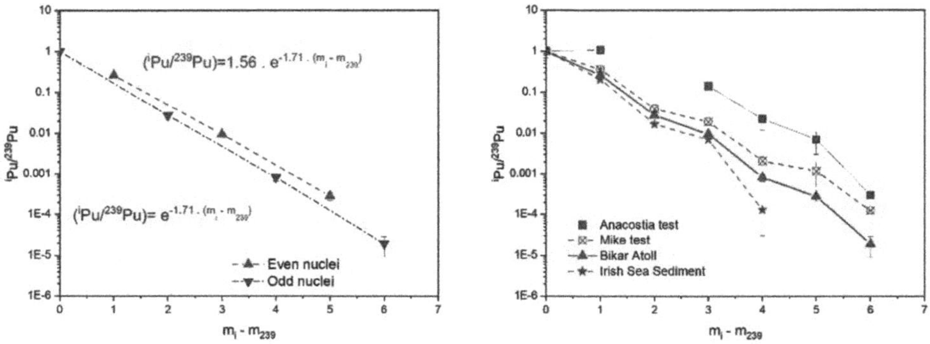

Figure 1 *a) Mean relative abundances of transuranium isotopes (in log scale), normalized to ^{239}Pu, in soil from Bikar Atoll. The even-odd effect is clearly observed. b) Comparison of Bikar data with data reported in other studies such as in debris from the Mike and Anacostia nuclear tests, and in sediment from the Irish Sea.*

4 CONCLUSION

^{239}Pu, ^{240}Pu, ^{241}Pu, ^{242}Pu, ^{244}Pu, ^{241}Am, ^{243}Am and ^{245}Cm were measured in soil samples from Bikar atoll, Marshall Islands, known to have been contaminated mainly by the Castle Bravo test. Data on the minor Pu and Am isotopes in environmental samples is scarce. The abundances of the transuranium radionuclides showed a consistent decreasing trend with the increase of the isotope mass, as expected for this type of event. Hence, the use of an exponential function to approximate the yield of heavy radionuclides in thermonuclear explosions with increasing mass seems to be a valid approach for estimating the age of a contamination. The radionuclide yield data also showed the even-odd anomaly of heavy nuclei as it has been observed in other thermonuclear explosions.

5 REFERENCES

1. Combined Radiochemistry Group, *Physical Review*, 1966, **148(3)**, 1192.
2. H. Diamond et al., *Physical Review*, 1960, **119(6)**, 2000.
3. G. Seaborg, *Proceedings of the National Academy of Sciences*, 1959, **45(4)**, 471.
4. A. Ghiorso et al., *Physical Review*, 1955, **99(3)**, 1048.
5. Los Alamos Radiochemistry Group, *Physical Review Letters*, 1965, **14(23)**, 962.
6. D. Dorn, *Physical Review*, 1962, **126(2)**, 693.
7. D. Dorn and R. Hoff, *Physical Review Letters*, 1965, **14(12)**, 440.
8. R. Hoff and D. Dorn, *Nuclear Science and Engineering*, 1964, **18(1),** 110.
9. Y. Lutostansky and V. Lyashuk, *KnE Energy*, 2018, **3(1)**, 1.
10. G. Bell, *Physical Review*, 1965, **139(5B),** B1207.

11. Y. Lutostansky, V. Lyashuk and I. Panov, Heavy and superheavy elements production in high intensive fluxes of explosive process. Cornell University, 2015.
12. Y. Lutostansky and V. Lyashuk, *JETP Letters*, 2018, **107(2)**, 79.
13. M. Smith-Norris, *The Journal of American-East Asian Relations*, 1997, **6(1)**, 1.
14. E. Hughes et al., *Proceedings of the National Academy of Sciences*, 2019, **116(31)**, 15420.
15. C. Hansen, US nuclear weapons : the secret history, 1988, Arlington, TX, New York, N.Y: Aerofax ; Distributed by Orion Books.
16. S. Hisamatsu and M. Sakanoue, *Health Physics*, 1978, **35(2)**, 301.
17. V. Noshkin et al., *Limnology and Oceanography*, 1975, **20(5),** 729.
18. V. Noshkin et al., Sediment studies at Bikini Atoll part 2. inventories of transuranium elements in surface sediments. 1997: United States.
19. V. Noshkin et al., Sediment studies at Bikini Atoll part 3. Inventories of some long-lived gamma-emitting radionuclides associated with lagoon surface sediments. 1997: United States.
20. J. Corcho Alvarado et al., *Journal of Environmental Radioactivity*, 2022, **242,** 106795.
21. J. Corcho-Alvarado et al., *Science of The Total Environment*, 2021, **801**, 149541.
22. S. Röllin et al., *Applied Radiation Isotopes*, 2009, **67(5),** 821.
23. H. Sahli et al., *Journal of Radioanalytical and Nuclear Chemistry*, 2017, **314(3)**, 2209.
24. P. Steier et al., *Nuclear Instruments and Methods in Physics Research Section B: Beam Interactions with Materials and Atoms*, 2013, **294,** 160.
25. P. Mitchell et al., *Applied Radiation and Isotopes*, 1998, **49(9)**, 1283.
26. V. Ageev et al., *Atomic Energy*, 1999, **87(5)**, 788.
27. A. Lashko, *Instruments and Experimental Techniques*, 2008, **51(3)**, 434.

CHAPTER 3 COINCIDENCE AND ANTI-COINCIDENCE GAMMA RAY SPECTROSCOPY IN RADIONUCLIDE IDENTIFICATION

S. Landsberger[a*], C. Egozi[a], W. Charlton[a], N. Kaitschuck[a] and F. J. Martinez[b]

[a]University of Texas at Austin, Nuclear Engineering Teach Lab, Pickle Research Campus, Building 159, Austin, Texas, USA, 78712, [b]Los Alamos National Laboratory, P.O. Box 1663, Los Alamos, New Mexico, USA, 87545

*s.landsberger@mail.utexas.edu

1 INTRODUCTION

Techniques for low-level radioactivity counting for alpha and beta particles, as well as gamma rays, have been used for many decades. The ability to make low-level counting successful has been a focus in many areas including nuclear engineering, nuclear physics, environmental monitoring, etc. Some examples of the use of low-level counting include the effect of uranium mining and remediation in the environment,[1] the determination of ^{241}Pu in environmental samples,[2] and radioactivity in food products.[3] Different techniques for low-level counting have been explored and implemented over the past few decades with varying levels of success. One of the primary issues in low-level radiation counting in gamma ray spectroscopy comes from the radiation emitted by natural radionuclides in the detector crystal and construction materials of the gamma ray detector.[4] Because this cannot be changed easily, the greatest area for improvement is with the control of external background radiation. This is most often done by implementing shielding techniques for reducing background radiation. This is implemented via the installation of shielding (typically lead, but also steel and tungsten) around the experimental setup to try to minimize external radiation making it to the detectors. Another method for reducing background radiation for low-level radiation counting has been the design of underground laboratories. These laboratories have been constructed and installed to try to eliminate all background radiation, including cosmic radiation.[5]

This technique is often coupled with the addition of external shielding for the counters and other external counting equipment. This technique has proven to be extremely effective with the background radiation count typically reduced by upwards of two orders of magnitude when compared to a similar above-ground configuration. The largest challenge for these types of laboratories is the cost and difficulty of designing and constructing the laboratory.

There are other techniques for achieving better signal to noise ratios for gamma ray spectroscopy, such as larger sample sizes and longer counting times for environmental radioactivity, longer irradiation and counting times for neutron activation analysis, and more efficient detectors. However, the main goal of our work in the past two decades has been to eliminate as much as possible spectral interferences and lower background radiation using coincidence, and anti-coincidence methods in gamma ray spectrometry. Several examples will be given of the powerful techniques in coincidence and anti-coincidence (Compton suppression) methods. More technical details of the results are available in the referenced papers.

1.1 Gamma-gamma coincidence

One of the very first attempts at the implementation of coincidence measurements was performed in 1924 by Walter Bothe and Hans Geiger.[6] They utilized the technique to validate that a recoil electron is produced simultaneously with a gamma ray during Compton scattering. Two newly developed Geiger needle counters separated with a very thin window of aluminium foil were utilized. When an isotope decays and emits gamma rays, the gamma ray emissions can occur either in coincidence or not in coincidence. When an isotope decays to one energy level by emitting a characteristic gamma ray and then subsequently decays to another energy level, emitting a second gamma ray, these gamma emissions occur within a very small window of time (except for few metastable states). If the decays of the gamma rays occur within a very short time window, nanoseconds or less, they are considered to be "in coincidence" with one another. To explain the concept of gamma-gamma coincidence, some sample gamma ray spectra will be examined. This concept can be easily understood with the use of sample gamma ray spectra to visually demonstrate what constitutes gamma-gamma coincidence. Basic gamma-gamma coincidence has been previously described by Adams and Dams.[7] The three possibilities of gamma-gamma coincidence are shown in Figure 1. The decay scheme[8] of ^{60}Co depicts two strong gamma rays in coincidence, while the one for ^{95}Zr has no gamma rays in coincidence and finally the one for ^{59}Fe shows gamma rays in coincidence and not in coincidence. Thus, a clear understanding of the decay schemes along with branching ratios is imperative to properly assess the feasibility of using the γ-γ coincidence for specific isotopes.

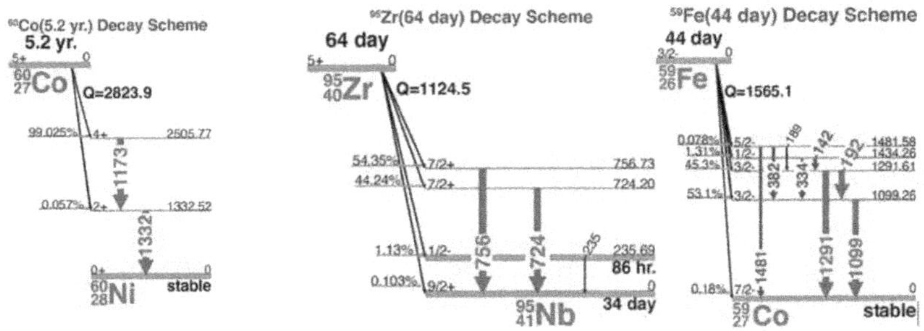

Figure 1 *Three decay schemes showing coincidence, non-coincidence, and a mixture of both.*

A typical set up of γ-γ coincidence is shown in Figure 2.

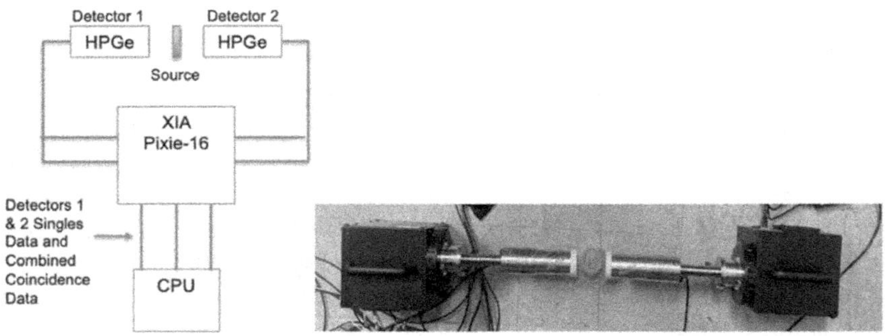

Figure 2 *Typical set up of γ-γ coincidence including two electronically cooled detectors and digital electronics. Reproduced from ref. 9 with permission from RAD.*

Several examples of the use of γ-γ coincidence are shown in the following figures. One of the unanticipated results of γ-γ coincidence is that virtually no shielding is required. This effect is illustrated by Figure 3 showing a spectrum of room background radiation in the singles and coincidence modes.[9] As can be seen the background is vastly reduced especially in the first 500 keV, as the Compton interaction reduces as the photon energy increases due to pair production above 1.022 keV.

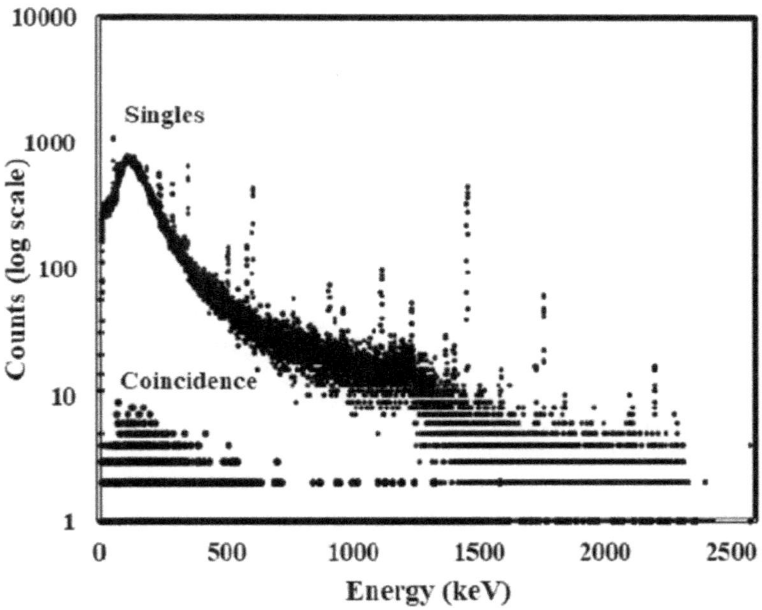

Figure 3 *Comparison of background radiation in single and γ-γ coincidence modes. Reproduced from ref. 9 with permission from RAD.*

The application of gamma-gamma coincidence for ^{239}Pu is shown in Figure 4.

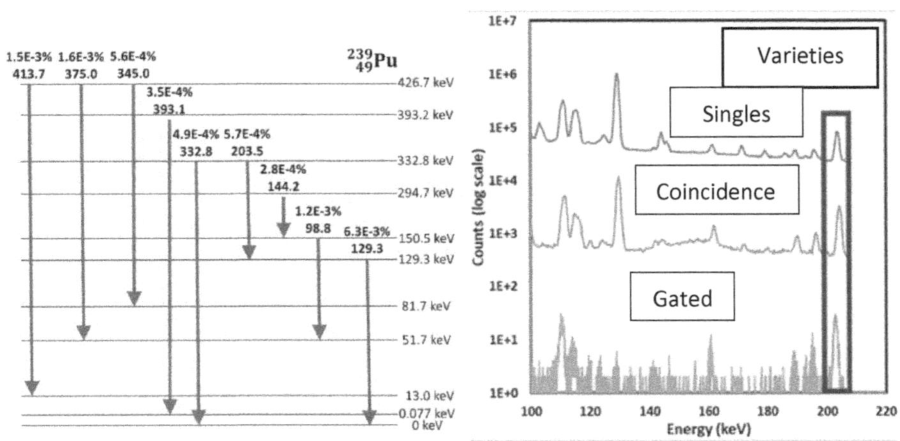

Figure 4 *The decay scheme of ^{239}Pu and the gating of the 129.3 keV gamma ray showing the 203.5 keV gamma ray on a virtually zero background using ^{239}Pu foil electrodeposited on nickel with an activity of 1.81 x 10^{6} Bq.*[9] *Reproduced from ref. 9 with permission from RAD.*

For the determination of arsenic in a rice flour reference material is seen in Figure 5.

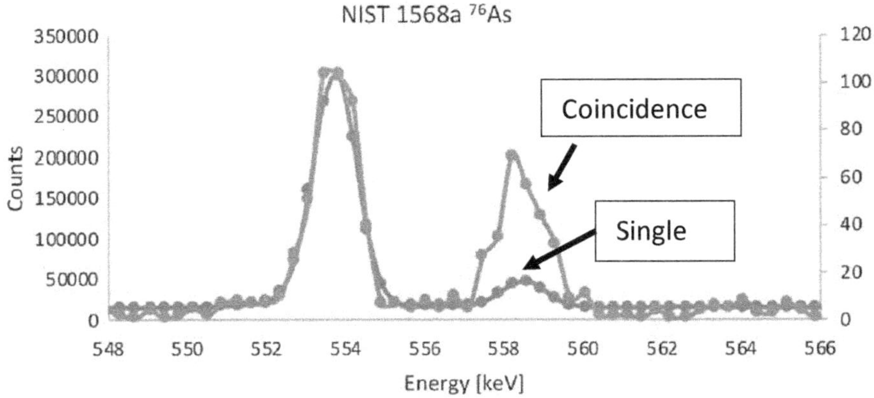

Figure 5 *The 559 keV emission from ^{75}As(n, γ)^{76}As reaction is shown alongside the 554.5 keV emission from ^{82}Br for both single detector and coincidence gate (on 657 keV) measurements of NIST rice flour 1568a.The use of a coincidence gate improves the strength relative to background, aiding in reducing the uncertainty in ^{76}As identification. Reproduced from ref. 10 with permission from Elsevier, Copyright 2022.*

In standard neutron activation analysis research, the 604.7 keV gamma ray for ^{134}Cs is the preferred photon as shown in Figure 6.[11]

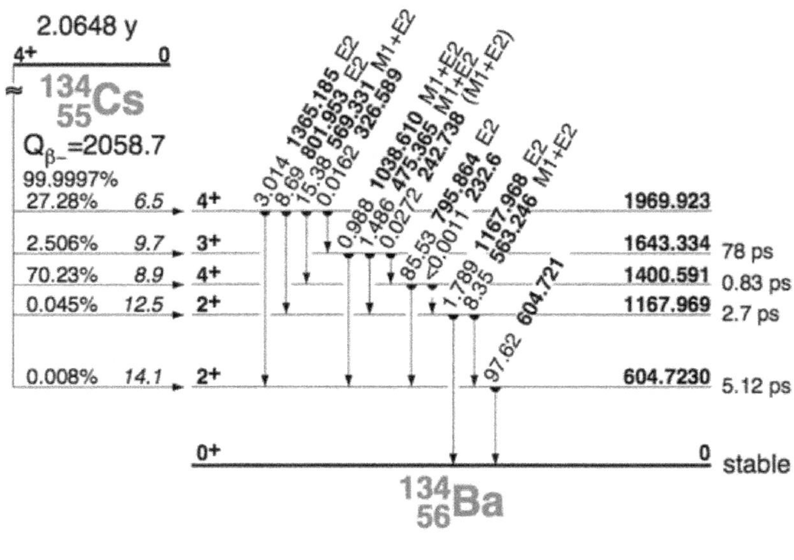

Figure 6 *Decay scheme of ^{134}Cs.*

However, there is a potential interference due to the 602.7 keV photon gamma ray from ^{124}Sb which has a 60-day half-life. However, ^{134}Cs was positively identified in NIST 2710a soil reference material, which also contains ^{124}Sb, utilizing an energy gate at 795.9 keV as the 569.3 keV and 604.7 keV gamma ray peaks are visible as seen in Figure 7. Thus, the spectral interference of the 602.7 keV gamma ray is eliminated.

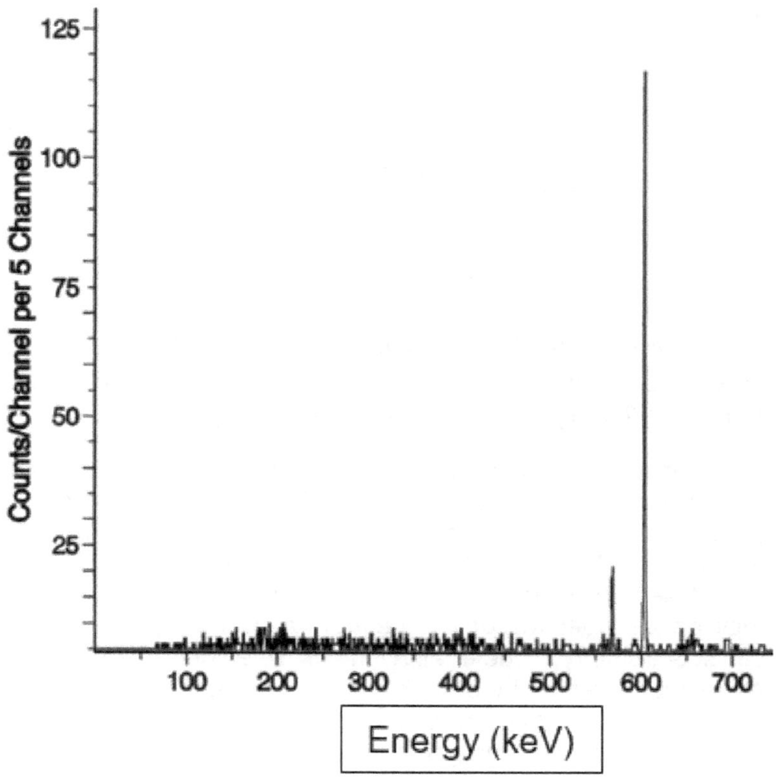

Figure 7 *Energy gate of 795.keV gamma ray showing the coincidence peaks of 569.3 keV and 604.7 keV gamma ray peaks. The $^{133}Cs(n,\gamma)^{134}Cs$ reaction is used.*

Conversely, ^{124}Sb was identified in NIST sample 2710a, but required energy gating at 1691.0 keV. Typically, the 602.7 keV line is used in NAA but has potential interferences from ^{134}Cs at 604.7 keV. When gated on 1691.0 keV, the interference free 602.7 keV peak is visible as seen in Figure 8.

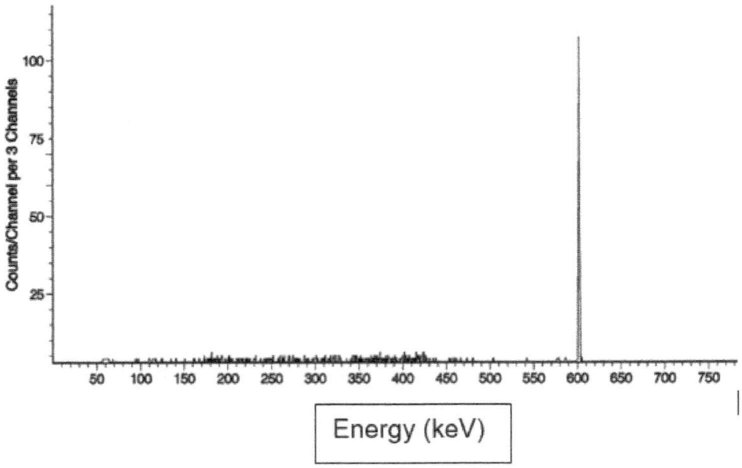

Figure 8 *Energy gate of 1691.0 keV gamma ray showing the coincident peak at 602.7 keV gamma ray. The $^{123}Sb(n,\gamma)^{124}Sb$ reaction is used. Reproduced from ref. 10 with permission from Elsevier, Copyright 2022.*

Other publications on γ-γ coincidence include quantifying radionuclide signatures from a γ-γ coincidence system in NORM,[12] coincidence gamma-gamma spectroscopy system for instrumental neutron activation analysis,[13] gamma-gamma coincidence spectrometer for instrumental neutron activation analysis,[14] multielement instrumental activation analysis based on gamma-gamma coincidence spectroscopy,[15] neutron activation analysis with γ-γ coincidence spectrometry,[16] determination of selenium in environmental standard reference materials by a γ-γ coincidence method,[17] digital gamma-gamma coincidence HPGe system for environmental analysis,[18] and neutron activation analysis using gamma-gamma coincidence to eliminate spectral interferences.[19]

1.2 Compton suppression (anti-coincidence)

The technique of Compton suppression has been utilized to reduce or eliminate the contribution of the Compton effect in gamma ray spectroscopy, in addition to the reduction of cosmic radiation, for low-level counting applications.[20-22] A Compton suppression system accounts for the contribution of the gamma ray, the scattering photon and recoil electron which are formed when the gamma ray interacts with the detector. A Compton suppression system is designed to independently measure both the singles spectrum as well as the spectrum where the scattering photons of the continuum have been suppressed. The basic interaction of Compton interaction is shown in Figure 9 for routine neutron activation analysis, while a typical system for routine neutron activation analysis is depicted in Figure 10.

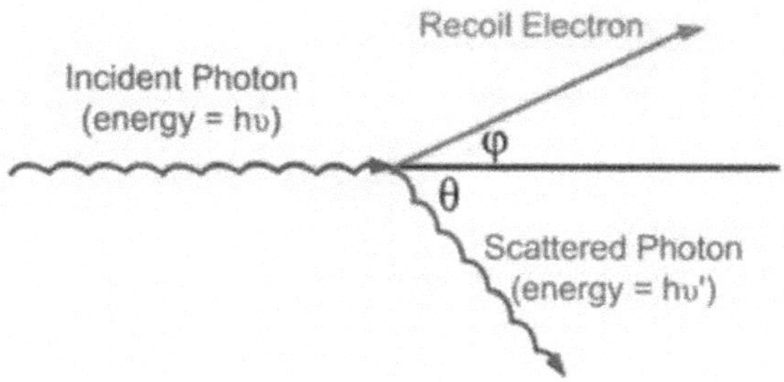

Figure 9 *Basic Compton interaction.*

Figure 10 *Cutaway view of Compton system for neutron activation analysis.*

Compton suppression is best exploited when the gamma ray of analytical interest is the only or major one that is involved in the beta decay process. For NAA some cases include ^{65}Zn, ^{51}Ti, ^{80}Br, ^{203}Hg, ^{198}Au, ^{52}Cr, $^{115}Cd/^{115}In$ and ^{128}I. Several examples of the use of Compton suppression for NAA and environmental radioactivity are shown in the following figures 11-17.

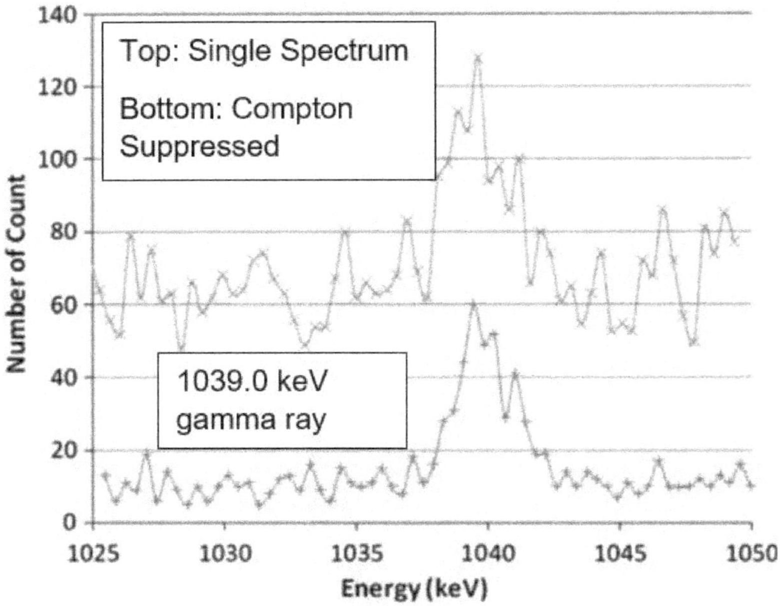

Figure 11 *Neutron activation of blueberries to determine copper through the $^{65}Cu(n,\gamma)^{66}Cu$ reaction using the 1039.0 keV gamma ray.*[23]

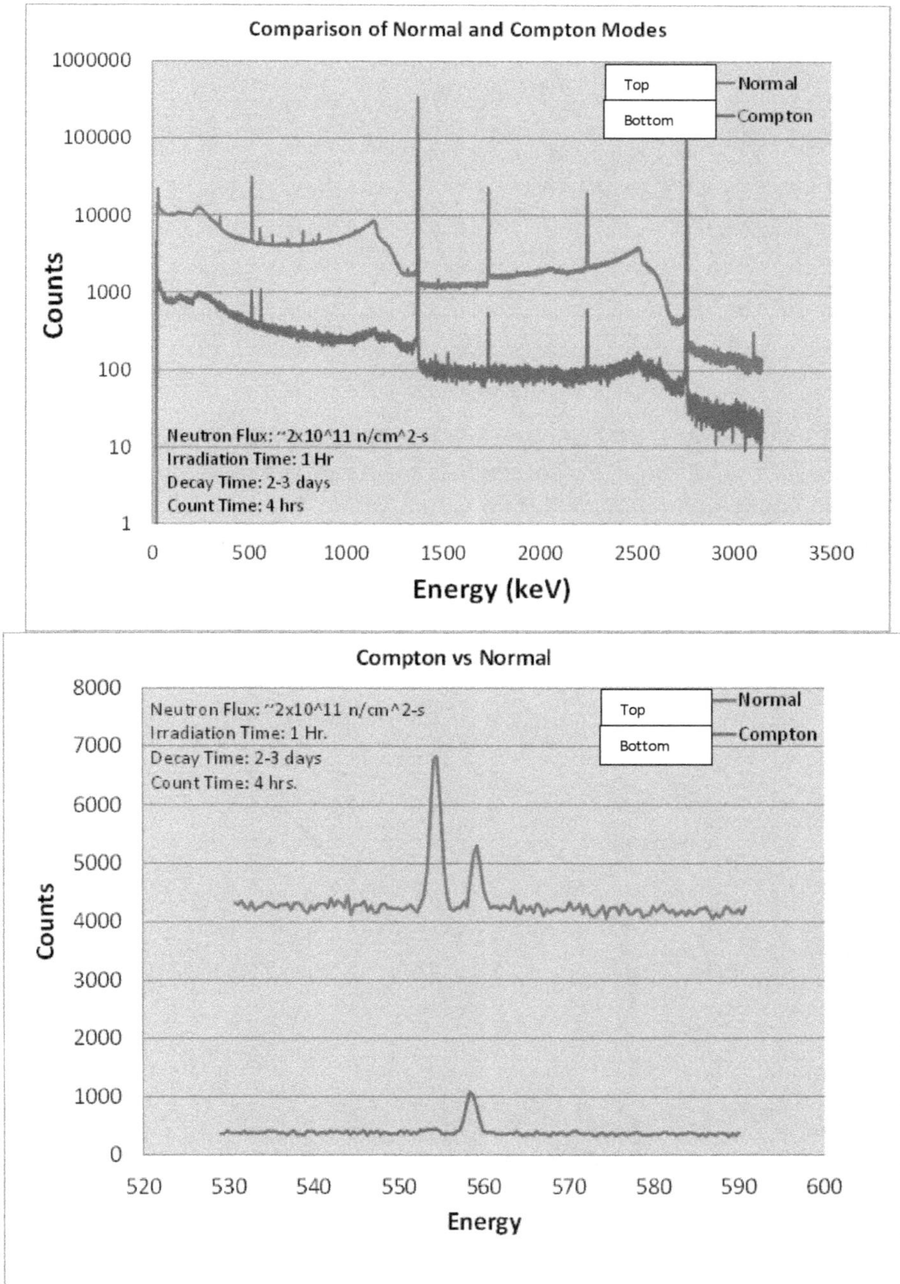

Figure 12 *Neutron activation analysis of water to determine arsenic through the $^{75}As(n,\gamma)^{76}As$ reaction using the 559.4 keV gamma ray. Top portion are the whole spectra with and without Compton suppression*[24]. *Bottom portion is the finer resolution of the spectra.*

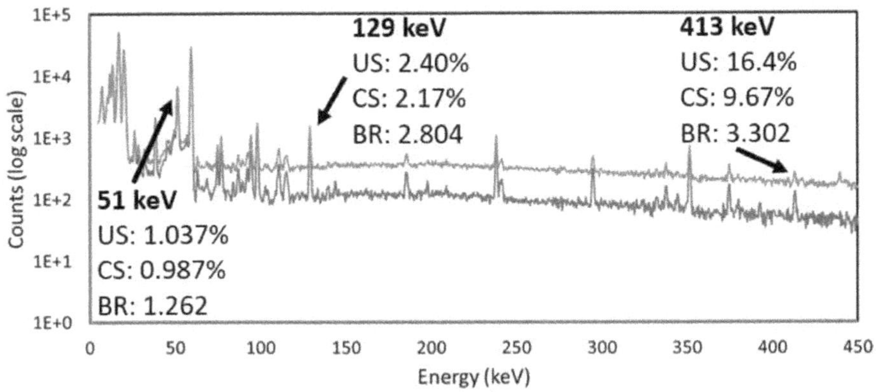

Figure 13 *Comparison of unsuppressed (US) and Compton suppressed (CS) spectra for a 3700 Bq* ^{239}*Pu source in a soil reference material. As can be seen the background reduction (BR) is reduced from 2.8 to 3.3 for the 129.3 keV and 413.y keV gamma rays. As expected in the region below 100 keV the Compton suppression is not as effective since the main interaction is the photoelectric effect.*[9]

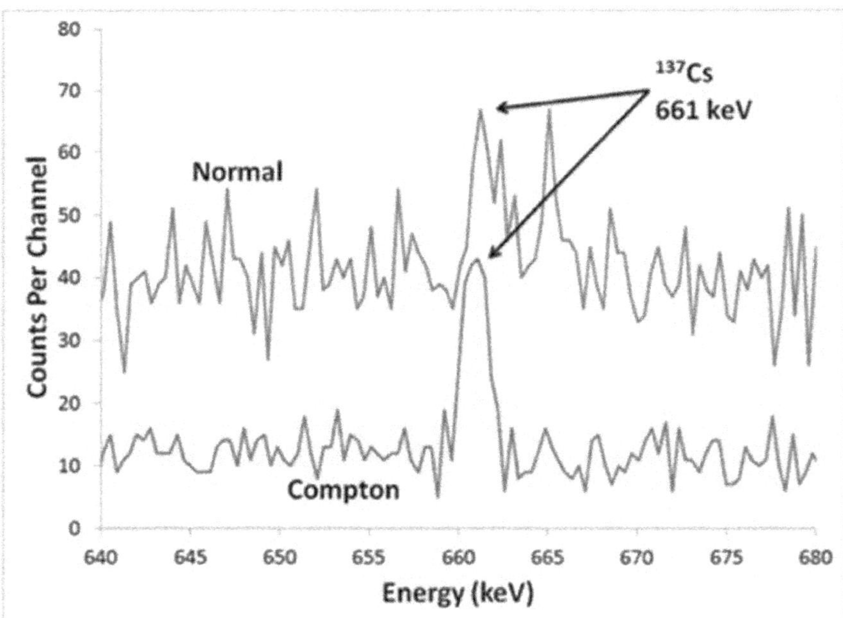

Figure 14 *Normal and Compton suppression spectra of* ^{137}*Cs in a 20-gram sample of soil counted for 24 hours. As can be seen the 661 keV photopeak is better formed and on a lower background giving a detection limit of less than ~0.7 Bq/kg.*[25]

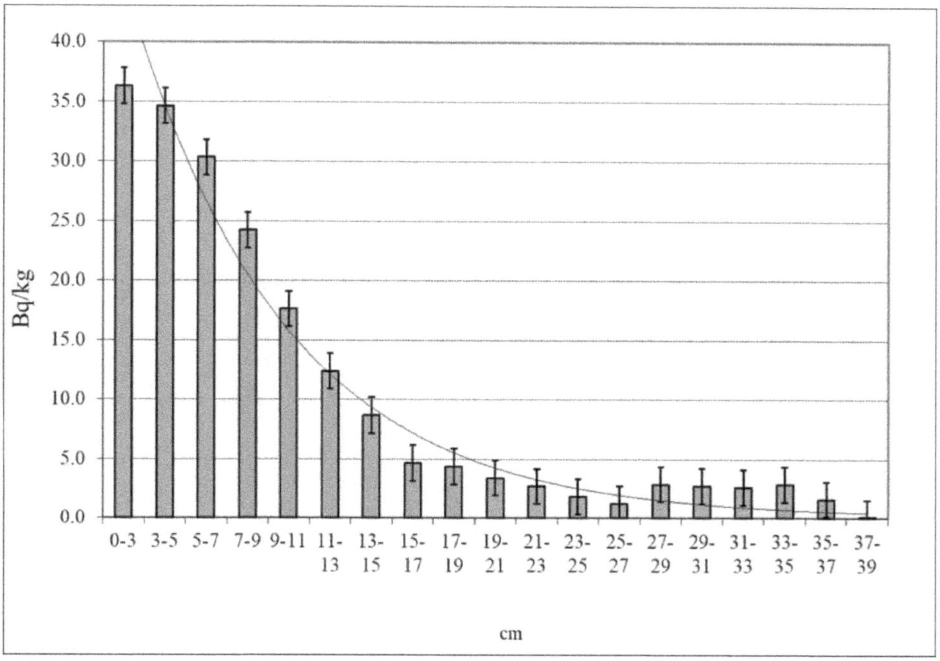

Figure 15 *The depth profile in soil using the Compton suppression of 661 keV photopeak of* ^{137}Cs.[25]

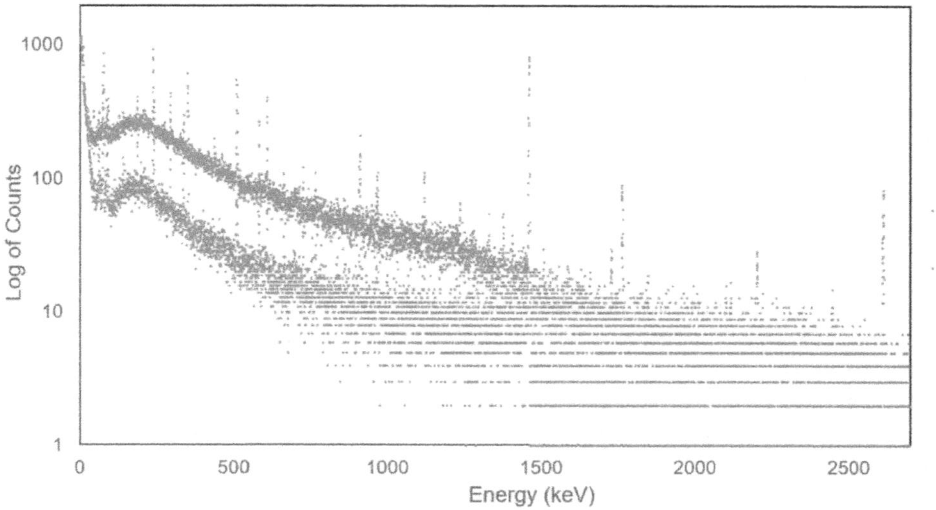

Figure 16 *Comparison a 20-gram soil sample counted for 18 hours for NORM measurements. Top spectrum is unsuppressed and bottom is Compton suppressed.*[25]

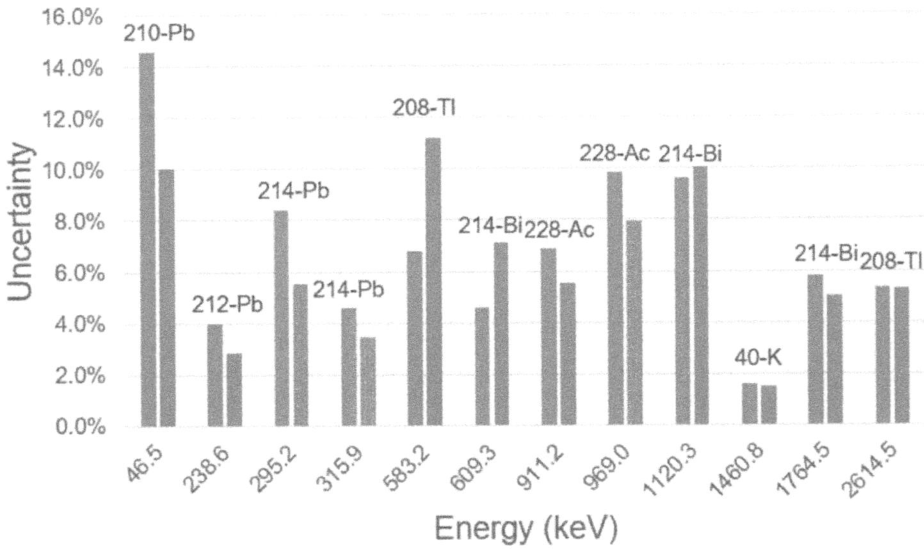

Figure 17 *Comparison of counting uncertainties of commonly used gamma rays in NORM measurements. Left bar is unsuppressed right bar is Compton suppressed. Gamma rays that are in coincident with others have higher uncertainties since Compton suppression is best exploited when the gamma ray of analytical interest is the only or major one that is involved in the beta decay process. As well Compton suppression has reduced effect for those photons of energies above 1000 keV since there is an increase in pair production above 1022 keV (twice the rest mass of the electron of 511 keV).*[25]

Other papers on Compton suppression include a comprehensive study for the determination of forty-eight elements in the certification of a hazardous standard reference material,[26] low level detection of praseodymium,[27] analysis of plant materials,[28] and minimizing sample sizes while achieving accurate elemental concentrations in neutron activation analysis of pottery.[29]

3 CONCLUSIONS

As with all traditional chemical analytical and radioanalytical methods, no one technique is a panacea for all elemental or isotopic constituents, respectively. However, with the advent of digital systems replacing complex electronic approaches and with much higher hyper-pure germanium detectors, coincidence, and anti-coincidence (Compton suppression) methods are now more than a small niche. There are now many more applications so these technologies in the field of gamma ray spectrometry including nuclear forensics. Nevertheless, a very careful evaluation of the specific decay schemes is needed for each stable or radioactive isotope to assure maximum likelihood of success employing the different methodologies used.

4 ACKNOWLEDGEMENTS

We acknowledge Los Alamos National Laboratory for their funding: LA-UR-23-30142.

5 REFERENCES

1. S. Niese, *Applied Radiation and Isotopes*, 1996, **47,** 1109.
2. M. Lee and C. Lee, *Radiochimica Acta,* 1991, **84,** 177.
3. L. Kaihola, Low Level Liquid Scintillation Counting of Radioactivity in Food Products, Application Note, www.perkinelmer.com https://resources.perkinelmer.com/corporate/pdfs/downloads/009647_01_app_radioactivityinfoodprod.pdf (accessed November 2022).
4. G. Knoll, (2010). *Radiation Detection and Measurement* (4th ed.). John Wiley & Sons Ltd.
5. S. Niese, M. Kohler and B. Gleisberg, *Journal of Radioanalytical Chemistry*, 1998, **233,** 167.
6. L. Bonolis, *American Journal of Physics,* 2011, **79,** 1133.
7. F. Adams and R. Dams, *Applied Gamma Ray Spectrometry, 1970,* Pergamon Press.
8. Idaho National Laboratory, *Gamma-Ray Spectrometry Catalog*, https://gammaray.inl.gov/SitePages/Home.aspx (accessed November 2022)
9. C. Egozi, F. Martinez, B. De Luna, J. Terry and S. Landsberger, "Preliminary Investigation for the Use of Digital Gamma-Gamma Coincidence Spectrometry to Detect ^{239}Pu, RAD Conference Proceedings, 2021, 5, 110, DOI: 10.21175/RadProc.2021.21
10. A. Drescher, M. Yoho and S. Landsberger, *Journal of Radioanalytical and Nuclear Chemistry*, 2018, **318,** 527.
11. https://blog.goo.ne.jp/takizawas/e/b05c0aebcad9c7daca51463b90a7236a (accessed November 2022).
12. R. Britton, M. J. Jackson and A, V. Davies, A. V., *Journal of Environmental Radioactivity*, 2015, **149,** 158.
13. J. Jakůbek*, et al., Nuclear Instruments and Methods in Physics Research, Section A: Accelerators, Spectrometers, Detectors and Associated Equipment*, 1998, **414,** 261.
14. B. Tomlin, R. Zeisler and R. Lindstrom*, Nuclear Instruments and Methods in Physics Research, Section A: Accelerators, Spectrometers, Detectors and Associated Equipment,* 2008, **589,** 243.
15. M. Vobecký, et al., *Analytica Chimica Acta*, 1999, **386,**181.
16. R. Zeisler, et al., *Journal of Radioanalytical and Nuclear Chemistry*, 2017, **314,** 513.
17. L. Wangen, E. Gladney and W. Hensley, *Analytical Chemistry,* 1980, **52,** 765.
18. N. Marković, P. Roos and S. Nielsen, *Applied Radiation and Isotopes,* 2017, **126**, 194.
19. F. Martinez and S. Landsberger, *Journal of Radioanalytical and Nuclear Chemistry*, 2022, https://doi.org/10.1007/s10967-022-08534-4.

20. S. Landsberger and S. Peshev, *Journal of Radioanalytical and Nuclear Chemistry,* 1996, **202,** 203.
21. S. Landsberger, *Journal of Radioanalytical and Nuclear Chemistry*, 1994, **179,** 67.
22. R. Britton, *Journal of Radioanalytical and Nuclear Chemistry*, 2012, **292,** 33
23. A. Michenaud-Rague, S. Robinson and S. Landsberger, 2012, **291**, 237.
24. S. G. Landsberger, G. George, S. Landsberger and G. Kuzmin, *Transactions of the American Nuclear Society*, 2014, **111,** 552.
25. Unpublished results at the University of Texas
26. *S. Landsberger and D. Wu, Journal of Radioanalytical and Nuclear Chemistry,* 1995, **193,** 49.
27. M. B. Stokley and S. Landsberger, *Journal of Radioanalytical and Nuclear Chemistry, 2018,* **318,** *369.*
28. M. A. Bacchi et al., *Journal of Radioanalytical and Nuclear Chemistry*, 2007, **271,** 345.
29. S. Landsberger and J. Yellin, *J. Archaeological Science: Reports*, 2018, **20,** 622.

CHAPTER 4 APPLICATION OF PLUTONIUM RADIOCHRONOMETRY TO UNDERSTAND MATERIAL PROCESSING IMPACTS FOR NUCLEAR FORENSICS

Matthew A. Higginson[a*], Christopher R. D. Gilligan[a], Brogen Dawkins[a] and Philip Kaye[a]

[a]AWE, Aldermaston, RG7 4PR, UK

*Matthew.Higginson@awe.co.uk

1 INTRODUCTION

Nuclear forensics analysis capabilities support legal investigation of nuclear and radiological materials found outside of regulatory control.[1] The field draws upon a wide range of analytical tools and techniques. This is with the aim of correlating measurable material parameters to the history of the unknown material to confidently answer investigative questions.[2,3]

Radiochronometry is a key measurement which aims to address the temporal questions posed within a forensic examination of a material. The measurement of trace decay progeny to determine model ages is well established.[4,5,6] Traceable measurements by Isotope Dilution Mass Spectrometry (ID-MS) are widely applied as they offer excellent accuracy and precision whilst minimising destructive sample utilisation.[7,8] Multiple decay pairs are analysed by forensic laboratories to build confidence that the model age represents a relevant date of purification. In order to underpin the use of these material parameters for assessment it is crucial to understand which process within the fuel cycle a model age is related to. This can be achieved by measurement of materials of known provenance from the nuclear fuel cycle.

Agreement between multiple chronometric pairs within measurement uncertainty is termed concordance. If concordant ages are observed in the measurement of multiple chronometric pairs this adds confidence in our assessment of the date of last chemical purification. It is also crucial to understand which specific fuel cycle processes have high separation factors to purify the decay products of interest to impart this signature. This allows assessment of which fuel cycle processes were likely to be associated with the model age measured. Disagreement between chronometric pairs is termed discordance and is also of high value for forensic analysis. Discordance has been shown to relate to the chemical fractionation or specific processing during the nuclear fuel cycle and gives nuanced information for interpretation.[9]

Nuclear forensic analysis relies on understanding the evolution of signatures during processing and a series of detailed studies are reported.[10, 11, 12] In order to build confidence in the interpretation of discordant chronometers in unknown materials it is crucial to have data within National Nuclear Forensic Libraries (NNFLs) of materials of known provenance, and processing history to support assessments.[13]

In the case of plutonium, we have previously reported methodologies for the flexible and robust measurement of a range of chronometer pairs. These methods are in line with those reported by other forensic laboratories.[14,15] We can measure traceably the $^{238}Pu/^{234}U$, $^{239}Pu/^{235}U$. $^{240}Pu/^{236}U$, $^{241}Pu/^{241}Am/^{237}Np$ and $^{242}Pu/^{238}U$ decay pairs by a combination of radiometric and mass spectrometric techniques. Efforts have also been supported to produce a reference standard for plutonium radiochronometry to underpin and validate methodologies.[14,15]

As plutonium decays, radiogenic helium is accumulated within a solid sample matrix. The chemical and kinetic behaviour of helium is sufficiently different than all other actinide series daughter products. A helium chronometric system may provide distinctive information regarding the history of the sample alongside isotopic information which is needed to assess the model decay within a given sample. As an inert gas with low solubility and high diffusivity, the Actinide-helium system is more likely to be reset during fuel cycle processing. The helium ages are likely to reflect the time of most recent material solidification or processing. Any discordance relative to other chronometer pairs (i.e. residual helium) could be used to infer temperature and time domain of production. Overall, this is more information to constrain the history of a material, but it may have limited application to small or young samples.

With a set of assumptions based on the measured isotopic profile and a closed system a model age can be estimated from the measured helium within a material. This material parameter may be useful to determine the date of last material solidification and heat treatment more accurately, but it is not widely applied for nuclear forensic analysis. We can melt a sample in an inert gas electrode furnace and measure the helium using a carrier gas through a glovebox breakthrough interface to a mass spectrometer. This technique is sensitive and will easily discriminate the 4 atomic mass unit from any relevant isobaric interferences whilst calibrant gases can be used to underpin measurement traceability.

In the processing of plutonium, the impact of wet chemical processing, material blending and the impact of thermal sample history are all relevant areas of interest for nuclear forensics. This is because they are common in almost all nuclear processing pathways. In order to better understand these processes samples of known provenance were produced and analysed by our reported methods to both test capability on realistic samples and to add data to the NNFL.

This paper aims to highlight analysis of samples of known provenance with the aim of better understanding the above impacts on plutonium radio-chronometric pairs.

2 EXPERIMENTAL

All chemicals used for radio-chronometry were of Optima grade (Fisher, Loughborough UK). Chemical yield tracers and reference materials were provided from calibrated stocks provided by NIST (USA) and NPL (Teddington, UK).

The separation and analytical methodology using Thermal Ionisation Mass Spectrometry (TIMS) and High-Resolution Gamma Spectroscopy (HRGS) are reported in detail in a prior publication.[11] Samples were dissolved gravimetrically and used for ID-MS measurements. Expanded uncertainties are reported at the k=2 confidence level unless otherwise stated. Isotopic and assay control materials and CRM-126a were analysed with each batch and were shown to be within the certified and consensus age value within measurement uncertainty.[14,15]

The helium measurements were carried out in an Eltra ONH800 electrode furnace coupled with to an RGA mass spectrometer calibrated with certified helium standards using a linearity calibration. A 0.5 g sample was melted and retained for further measurements, the helium from the sample was quantified and the model age was estimated based on the assumed material decay from the measured isotopics and assay by TIMS.

3 RESULTS AND DISCUSSION

The model age calculated should be relatable to the processing history of a nuclear material of known provenance. Two key processes within the fuel cycle were targeted with measurement of materials of known provenance are presented as case studies below.

Firstly, we aimed to understand the impacts of wet chemical processing by the Plutonium URanium EXtraction (PUREX) process on radiochronometric signatures. A material which was wet chemically purified by the PUREX process in June 1975 and further processed to an oxide and then metal in 1978 was chosen for this study.

Secondly, the impact of material blending, which is encountered in many production, recycle and fuel fabrication processes was investigated. A 50:50

mixture of metal materials with known ages was blended, cast and analysed to determine if composite feedstock ages are predictable in a sample of known provenance.

3.1 Plutonium oxide purified by wet chemistry

This sample was known to have been purified by solvent extraction after reprocessing in 1975. The material was further processed in 1978 and archived leaving a sample for analysis. The PUREX flowsheet has high separation factors for U, Am and Np and we would predict complete purification. The results are shown below (Figure 1).

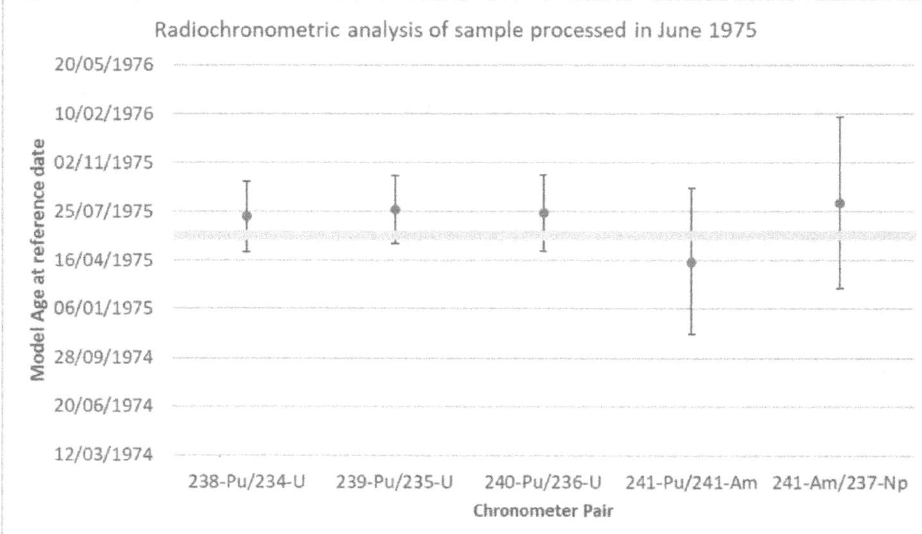

Figure 1 *Model age of plutonium material measured in this study; uncertainties are reported at k = 2. The bar indicates the known time of the last purification process in 1975.*

As can be observed, the model ages are all in agreement with the known history of the material. In this case the model ages report the last date of chemical purification and not the calcination of the material to an oxide in subsequent years. This is useful information to support forensic assessments. This also agrees with previous studies looking at analogous uranium processing by wet chemical methods.[14]

3.2 Analysis of mixed feedstock plutonium materials by radiochronometry

A 50:50 mixture of two plutonium metal materials was chosen for this study to investigate the impacts of blending materials on the apparent model ages. A plutonium material produced in 2000 from recycled older material was combined with a material from 1992 and cast by vacuum induction melting on 26/06/2001 to produce a metal sample. The material was cast in 2001 but the calculated mean model age based on the feedstocks is Jan-1997 relative to the last date of feedstock purification.

The radiochronometric measurements of the sample are shown above (Figure 2). It can be observed that the uranium chronometers preserve the composite feed age of the two blended materials which is key forensic information. The Am and Np chronometer are discordant with the uranium chronometers and in this case it is likely to be an additional signature that is retained relating to prior incomplete processing of the recycled materials by pyro-processing prior to metal production. The model ages agree with the known provenance of the material and reflect a combination of blending and process separation efficiency of the material. The complexity of the material however leads to a model age that does not reflect the last relevant processing stage of the material (i.e. the date of the casting of these samples in June 2001). However additional forensic signatures may be applicable to provide this information.

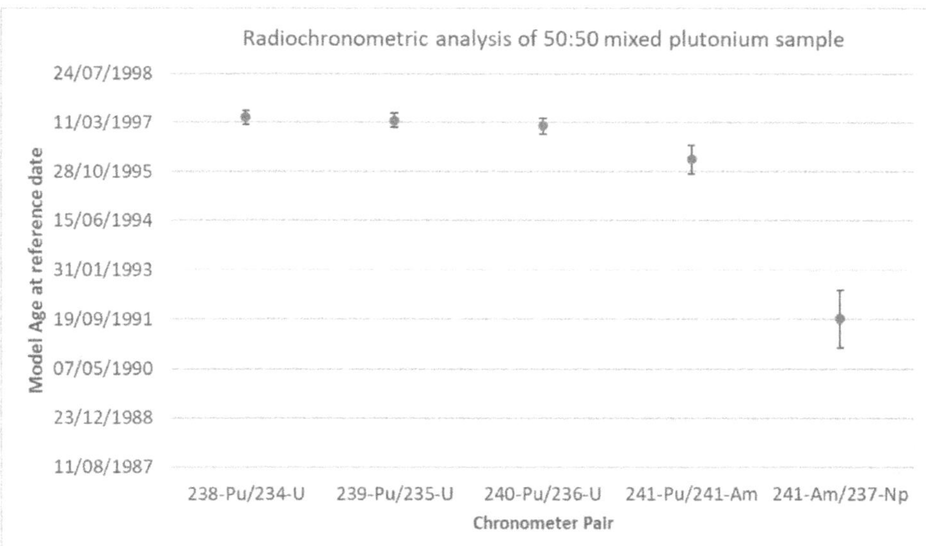

Figure 2 *Model age of plutonium material measured in this study, uncertainties are reported at k = 2. The U chronometer pairs are in agreement with the feedstock mean model age of Jan 1997, whereas the Np and Am chronometer are discordant, and are an artefact of the prior processing history of the recycled material.*

Application of the helium chronometer is predicted to be more applicable to this type of sample. The sample was measured by evolution of the helium by heating of the sample in a furnace followed by measurement by gas mass spectrometry traceable to helium standards. The model age from the helium chronometer was determined to be 14/02/2001 +/- 172 days (k = 2). This model age is in agreement with the known casting date of the material in June 2001, showing the utility of this chronometer for this sample type as well as the utility of the Pu decay series chronometer pairs to highlight feedstock blending ages in materials. Further studies are needed to underpin the use of this tool for nuclear forensics, relating to the thermal processing of samples and better understanding the diffusivity of helium in solid plutonium matrices.

This type of sample which contains real world complexity would also benefit from the wider application of other forensic tools such as Secondary Ionisation Mass Spectrometry (SIMS), Microscopy and Metallurgical analysis to obtain spatial and metallurgical information of the material. In the case of the radiochronometric signature utility, firstly we can bound a minimum sample age from these measurements and use discordance to inform a nuanced sample processing history relating to prior processing of the material.

Overall, the analysis of this material of known provenance yielded excellent information to inform assessments and add knowledge on fuel cycle impacts to the NNFL.

5 CONCLUSIONS

We have successfully applied our validated plutonium radiochronometry methods to two samples of known provenance to add to our understanding of fuel cycle impacts on apparent model ages for nuclear forensics.

The model age via the Pu/U and Pu/Am/Np decay pairs has been shown to accurately predict the known production history in plutonium samples which were wet chemically processed by the PUREX process.

In the case of a more complex metal sample produced by the blending of two feedstocks, measurement of a suite of decay products, including the use of the He/An chronometer pair, allowed for a nuanced understanding of the material. The composite feed age was retained in the Pu/U chronometer which is essential information to understand when applying radiochronometry to these complex sample types. The He/An chronometer was shown to be useful in determining the last date of material solidification. As a set of measurements, we are able to correlate these measurable parameters to the production history of the material to answer the temporal questions on the sample history. Further controlled studies are required to further develop these techniques for nuclear forensic applications.

6 CONFLICTS OF INTEREST

There are no conflicts to declare.

7 ACKNOWLEDGEMENTS

The research herein was funded and supported by AWE.

8 REFERENCES

1 F. Taylor, M. Higginson, O. Marsden and J. Schwantes, *J Radioanal. Nucl. Chem.,* 2020, **323**, 415.

2 M. Kristo , A. Gaffney N. Marks, K. Knight, W. Cassata and I. Hutcheon, *Annual Review of Earth and Planetary Sciences*, 2016, **44:1**, 555.

3 K. Mayer, M. Wallenius, I Ray, *Analyst*, 2005, **130**, 433.

4 K. Mayer, M. Wallenius, Z. Varga, *Chem. Rev.*, 2012, **113**, 884.

5 R. Williams and A. Gaffney, *Proc. Radiochim. Acta*, 2011, **1**, 31.

6 M. Higginson, P. Thompson, B. Dawkins, F. Taylor and P. Kaye, *Environ. Radiochemical Anal.*, 2020, **13**, ebook.

7 G. Eppich, R. Williams, A. Gaffney and K. Schorzman, *J Anal. At. Spectrom.*, 2013, **28**, 666.

8 M. Kristo, R. Williams, A. Gaffney, T. Kayzar-Boggs, K. Schorzman, P. Lagerkvist, A. Vesterlund, H. Ramebäck, A. Nelwamondo, D. Kotze, K. Song, S. Lim, S. Han, C. Lee, A. Okubo, D. Maloubier, D. Cardona, P. Samuleev, I. Dimayuga, Z. Varga, M. Wallenius, K. Mayer, E. Loi, E. Keegan, J. Harrison, S. Thiruvoth, F. Stanley, K. Spencer and L. Tandon, *J Radioanal. Nucl. Chem.* 2018, **315**, 425.

9 M. Higginson, T. Kayzar-Boggs and A. Gaffney, *J Radioanal. Nucl. Chem.,* 2022, **331,** 4799.

10 Nuclear Forensics International Technical Working Group (ITWG) Collaborative materials exchange exercise after action and lessons learned report 4. Coordinator: Schwantes J (29/10/2015) Pacific Northwest National Laboratory. PNNL-24410.

11 Nuclear Forensics International Technical Working Group (ITWG) Round Robin 3 Exercise after action and lessons learned report 3. Coordinator: r. Hanlen (25/04/2011), Pacific Northwest National Laboratory. PNNL-20079.

12 Z. Varga, M. Wallenius, M. Krachler, N. Rauff-Nisthar, L. Fongaro, A. Knott, A. Nicholl and K. Mayer, *TrAC Trends in Analytical Chemistry*, 2022, **146**, 116503.

13 A. Vesterlund, J. Canaday and D. Chamberlain, *J Radioanal. Nucl. Chem.*, 2022, **331**, 639.

14 M. Higginson, J. Dunne and C. Gilligan, *J Radioanal. Nucl. Chem.*, 2022, **331**, 3591.

15 K. Mathew, T. Kayzar-Boggs, Z. Varga, A. Gaffney, J. Denton, J. Fulwyler, K. Garduno, A. Gaunt, J. Inglis, R. Keller, W. Kinman, D. Labotka, E. Lujan, J. Maassen, T. Mastren, I. May, K. Mayer, A. Nicholl, C. Ottenfeld, T. Parsons-Davis, D. Porterfield, J. Rim, J. Rolison, F. Stanley, R. Steiner, L. Tandon, M. Thomas, R. Torres, K. Treinen, M. Wallenius, A. Wende, R. Williams and J. Wimpenny, *Analytical Chemistry*, 2019, **91**, 11643.

CHAPTER 5 APPLICATION OF RADIOCAESIUM MICROSCALE OBSERVATION METHODOLOGY TO PARMELIOID LICHEN AND ULTRASTRUCTURAL ANALYSES USING STEM-EDS

Terumi Dohi[a*], Kazuki Iijima[a], Masahiko Machida[b], Hiroya Suno[b,c], Yoshihito Ohmura[d], Kenso Fujiwara[a], Sigeru Kimura[e] and Futoshi Kanno[f]

[a]Sector of Fukushima Research and Development, JAEA, Fukushima, Japan, [b]Center for Computational Science & e-Systems, JAEA, Chiba, Japan, [c]Research Center for Computing and Multimedia Studies, Hosei University, Tokyo, Japan, [d]Department of Botany, National Museum of Nature and Science, Ibaraki, Japan, [e]Nuclear Engineering Co., Ltd., Ibaraki, Japan, [f]Pesco Co., Ltd., Tokyo, Japan.

*dohi.terumi@jaea.go.jp

1 INTRODUCTION

Lichens are well known for their ability to accumulate radiocaesium released into the atmosphere, by uptake in the thallus via wet and dry deposition over long periods of time. Thus, lichens are often used as biomonitors for radiocaesium contamination levels.[1] After the Fukushima Dai-ichi Nuclear Power Plant (FDNPP) accident in 2011, a number of foliose macro-lichens with layer-structures in their thalli (Figure 1) were analysed as potential biomonitors of radiocaesium contamination.[2]

In order to reveal exactly where radiocaesium is retained in the thallus tissues and in what chemical form, we developed a method for microscale observation of radiocaesium localisation in lichen thallus tissues using the species *Parmotrema tinctorum*.[3] Although we previously suggested Cs^+ ions can be retained in pigmented tissues of the lower cortex of *Parmotrema tinctorum*[3], it is not clear whether other species also have the same tendency. The purpose of this study was to examine the localisation of radiocaesium in other lichen species using *Punctelia borreri* as an example. In addition, we performed ultrastructural observations of the thallus tissue of *Parmotrema tinctorum* to investigate where fine particles might be present inside the tissue. For this purpose we used energy-dispersive X-ray spectrometers (EDS) with an ultra-high detection efficiency, interfaced to a scanning transmission electron microscope (STEM).

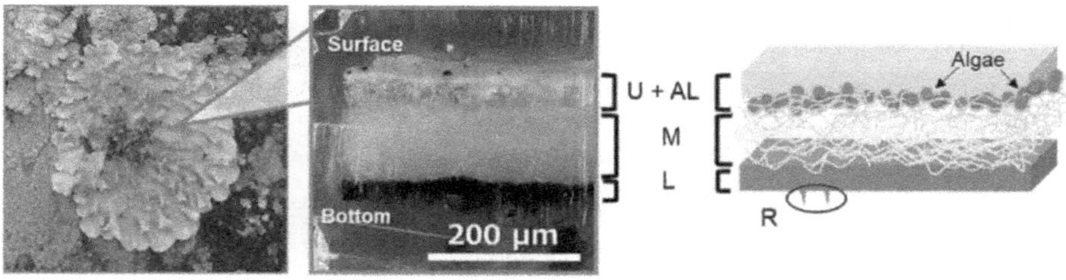

Figure 1 *Parmelioid lichen sample image. U + AL, upper cortex including the algal layer; M, medullary layer; L, lower cortex; R, rhizine (growth, density and shape depend on the species). Rhizine structures are akin to roots that attach the lichen onto the substrate on which it is growing (but they do not function like the root system of vascular plants).*

2 EXPERIMENTAL

2.1 Microscale observation of radiocaesium localisation in *Punctelia borreri*

For this investigation a sub-sample of the lichen *Punctelia borreri* was used. This species had been collected previously within Okuma town (1 km from the FDNPP) six years after the 2011 accident (and was also used in a previous study).[4] This species is the target parmelioid lichen (*Parmeliaceae*) used by JAEA as a biomonitor for studying radiocaesium fallout within the Fukushima prefecture.[2] The activity concentration of ^{137}Cs in this sample was ca. 3000 Bq g^{-1} (in 2017) and in addition, when analysed by autoradiography (Typhoon FLA7000, GE Healthcare) in a previous study,[4] the radiocaesium showed both uniform and spotted/clustered distributions in the thalli. In this study fragments containing a uniform distribution of radiocaesium were used.

The method for microscale observation of radiocaesium localisation in lichen tissue described here is the same as the method developed in our previous study,[3] where it is discussed in more detail. Here, the procedure is summarised in Figure 2. A single thallus sample taken from *Punctelia borreri* was dissected using a razor into smaller sized fragments, circa 2 mm by 5 mm. The fragments were exposed for 24 hours to an imaging plate (IP) to allow imaging of the radiocaesium distribution (Figure 3). After the fragments were fixed and embedded in EPOK resin, the specimens were sliced into 5 μm-thick sections, from the thallus surface (upper cortex) through to the bottom side (lower cortex including some of the rhizines (but not all)) using an ultramicrotome (Leica EM UC7, Leica Microsystems). The sliced sections were individually laid out on a plastic cell plate in 4 rows ("a" to "d") and 18 columns ("1" to "18") in order to clearly position each section of thallus tissue (e.g. a-1 layer).

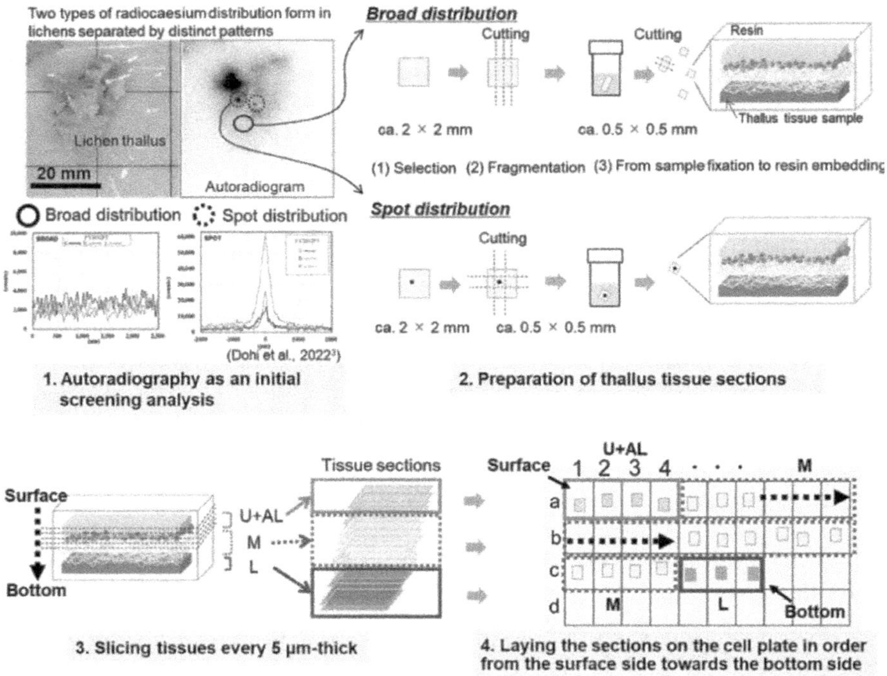

Figure 2 *Schematic diagram of the sample preparation process for microscale observation of radiocaesium localisation in lichen thallus tissue.*

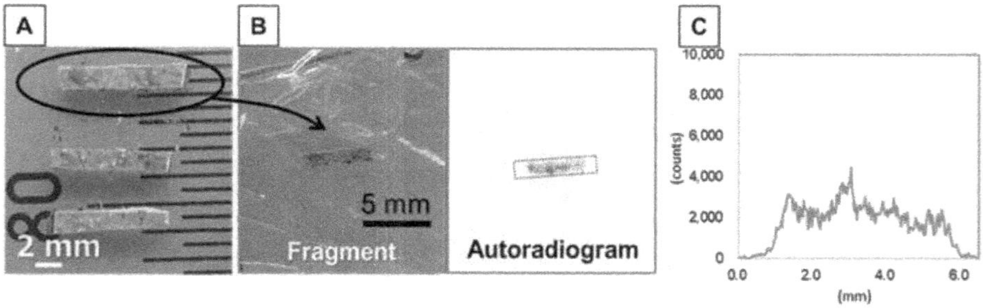

Figure 3 *Radiocaesium distribution in a thallus fragment of Punctelia borreri before the preparation of tissue sections. (A) Photo image of fragments. (B) Image and autoradiogram of fragment sample. (C) Signal intensity pattern of autoradiography within the range indicated by the dotted box in (B). The x-axis shows the linear distance of the distribution (mm), the y-axis shows the signal intensity value (counts).*

Once positioned, the sections of thallus tissue were observed using a digital microscope (VHX-8000, Keyence) to allow identification of individual tissue parts. Next, the sliced sections were exposed to the IP for 12 days inside lead shielding (to reduce any effects from background radiation) (Figure 4). After exposure was complete, the method of microscale observation developed previously,[3] was implemented in order to examine the radiocaesium distribution within each of the sections. The radiocaesium intensity was determined by subtracting the background signal (intensity) in the autoradiogram.

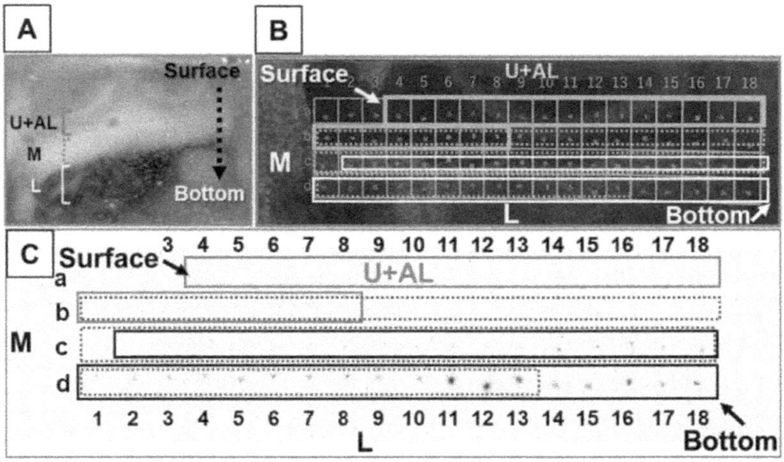

Figure 4 *Radiocaesium distribution in thallus sections of Punctelia borreri collected six years after the Fukushima accident (2011). (A) Cross-sectional image of the thallus tissue embedded in resin before microtome slicing. U+AL, upper cortex including the algal layer; M, medullary layer; L, lower cortex layer including some of the rhizines. (B) The tissue sections on a cell plate. Frames are sections containing tissue. Grey, dotted line and white (or black in (C)) boxes show the areas corresponding to the U+AL, M, and L, respectively. Overlapping boxes show that two layers were contained in the frames. (C) Autoradiograph images of each tissue section.*

2.2 Observation and analysis of ultrastructure in the cortex of *Parmotrema tinctorum*

Before undertaking examination of lichens containing fallout radiocaesium (approximately micron- and sub-micron sized particles containing radiocaesium) in future studies, it was decided to test uncontaminated samples of the same species of lichen, in order to ascertain the approximate position of fine particles trapped on the thallus surface within the lichen cortical tissue. For this study we used a sample of the lichen *Parmotrema tinctorum* that was collected in Sep. 2016 in the Gifu prefecture in Japan.[3] Due to the large distance between where the sample was collected in the Gifu prefecture and the FDNPP, this sample was thought to be unaffected (or at least, not significantly affected) by the accident fallout and was therefore considered suitable for use as a "background" sample.

The sample was fragmented and fixed in the same way as described above and in our previous study.[3] The fragmented samples then went through a number of sequential pre-treatments:

- Immersed with 1% osmic acid in 100 mM sodium phosphate buffer solution (pH 8.0) for 1 hour.
- Immersed three times in 100 mM sodium phosphate buffer solution for 10 minutes each time at room temperature.
- Washed with ultrapure water.
- Dehydrated in 100% ethanol and left overnight (approximately 16 hours).
- Immersed in propylene oxide (PO) three times for 10 minutes each time at room temperature.
- Sequentially immersed in solutions with different ratios of PO and prepared EPOK resin, as follows: PO: resin (2:1) for 24 hours, (1:1) for 24 hours, (1:2) for 72 hours.
- Placed in resin for 48 hours.

Finally, the resin-embedded specimens were prepared according to the same procedure for embedding in resin for section preparation as in the previous study.[3] The specimens were sliced into 200 nm-thick sections to allow observation of the cross-sectional structure of the thallus tissue. The ultrathin sections were stained with lead citrate staining solution,[5] to provide contrast against the electron beam for TEM analyses.

The ultrastructure of the thin cross-sections was observed, and elemental map analyses were also carried out, using a STEM-EDS (JEM-2800, JEOL) at an acceleration voltage of 100 kV.

3 RESULTS AND DISCUSSION

3.1 Microscale observation of radiocaesium localisation in *Punctelia borreri*

Sections of *Punctelia borreri* thallus tissue that showed broad radiocaesium distribution were prepared for analysis. Autoradiography was used to observe the radiocaesium distribution in each of the 5 µm-thick sections, from the upper cortex to the lower cortex including the rhizines. Radiocaesium was detected in the c-8 to d-18 layers, i.e. around the mixed medullary layer and in the lower cortex layer including the rhizines (Figure 4). These sections all contained brown pigmented tissue (often referred to as "melanin-like substances"),[3,6] consistent with our previous study.[3] Digital microscope observation indicated what appeared to be a mixture of both the lower cortex and rhizines contained within the d-1 to d-18 layers. The rhizines are formed by the extension of the cortex tissues,[7] and the brown pigmented tissue observed within the rhizines was also assumed to be melanin-like substances. Thus, radiocaesium was found to be distributed in the pigmented tissues of not only the lower cortex but also that of the rhizines. This result is consistent with the findings of our previous study,[3] and revealed that radiocaesium was located in the brown pigment of the species *Parmotrema tinctorum* and *Punctelia borreri*.

3.2 Observation and analyses of ultrastructure in the cortex of *Parmotrema tinctorum*

Both the upper and lower cortex appeared to have been conglutinated between the hyphae in the specimens of *Parmotrema tinctorum*. This finding was similar to that of the ultrastructure of other Parmeliaceae lichens in other studies, such as *Parmelia tiliacea*[8] and *Cetrelia olivetoruml.*[9] In addition, sub-micron particles were found to be buried in the surface tissue of the upper cortex (Figure 5). In a STEM bright-field (BF) image, particles appear dark because of their higher densities than surrounding tissue regions that appear bright. The sub-micron particles contained O, Si, Al, Mg and Fe, thus suggesting that they might be aluminosilicate particles. Furthermore, sub-micron particle was found at depth of ca. 2 μm from the surface of the upper cortex in the brightest regions of the STEM-BF image (which appeared to be areas of aggregation) (Figure 6). The aggregations, which looked like the boundaries of extracellular matrices (ECMs), were presumed to be crystalline secondary metabolic products (e.g. atranorin) dissolved during the fixation and/or dehydration of the specimen.[8-10] Fe and O detected in the particle, suggested that it was an iron-oxide particle (Figure 6). In our previous study,[3] observations of the thallus surface using scanning electron microscopy showed that micron-sized particles were partially buried in the surface tissue of *Parmotrema tinctorum*. It was considered that the particles may be embedded in the surface tissue[3] which arise from the ECMs and secondary metabolic products in the hyphae cell wall.[10] In addition, radiocaesium containing particles were detected around the medullary layers in the tissue sections of *Parmotrema* tinctorum.[3] Our results show that the particle we detected (Figure 6) was present inside the upper cortex tissue within what were presumed to be metabolite products. Moreover, the particle did not appear to have penetrated into the inside of the tissue through any cracks on the surface of the upper cortex, as none were observed around the particle that led to the surface (Figure 6). This indicates that particles could be retained on the inside surface tissue by the development of the cortex matrix including secondary products over a long period of time, although further analyses will be required to confirm this. Thus, radiocaesium particles originally trapped on the lichen surface would also be present inside the cortex tissue in the same way.

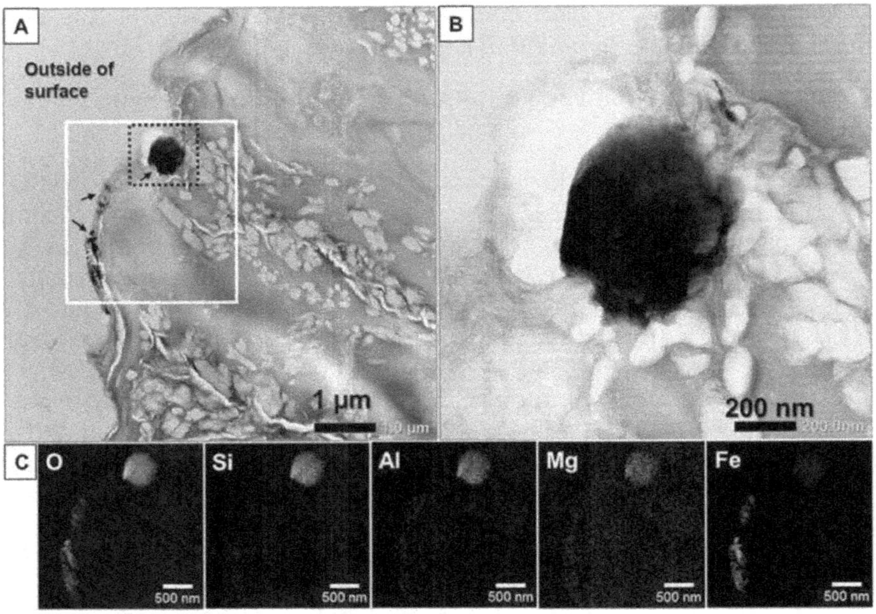

Figure 5 *(A) STEM-BF image of a vertical cross-section of the upper cortex of a sample of Parmotrema tinctorum. Arrows show sub-micron particles. (B) A magnified image of the dotted-line boxed area in (A). (C) Elemental maps of the white-boxed area in (A) as analysed by STEM-EDS.*

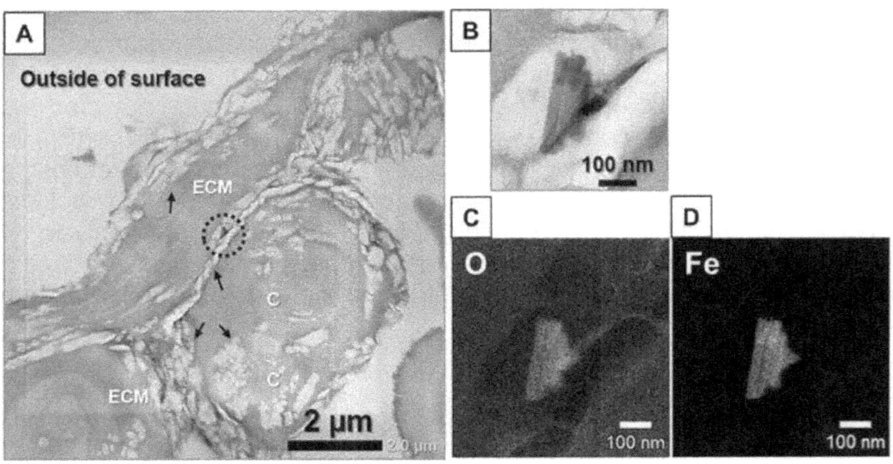

Figure 6 *(A) STEM-BF image of a vertical cross-section of the upper cortex of a sample of Parmotrema tinctorum. Arrows point to crystalline secondary products that dissolved during sample preparation; ECM, extracellular matrix; C, cortical cell. (B) Sub-micron particle image enclosed by the dotted line in (A). (C), (D) Elemental distribution of iron (C) and oxygen (D) in the image seen in (B) by STEM-EDS analysis.*

4 CONCLUSIONS

In this study, our novel methodology[3] for microscale observations of radiocaesium in lichen was applied to the lichen species, *Punctelia borreri*. Localisation of radiocaesium in the parmelioid lichen was consistent with that found in the species, *Parmotrema tinctorum* and *Punctelia borreri*. Ultrastructure observations of the upper cortex tissue in *Parmotrema tinctorum* confirmed a sub-micron particle was retained inside the cortical tissue. The presence of radiocaesium particles in cortical tissue, is likely the same as for the presence of other sub-micron particles in cortical tissue, and may be a mechanism that contributes to longer term retention. The methodological approaches presented here are expected to contribute to future studies on the mechanisms of radiocaesium accumulation not only in lichens but also in other biological organisms such as wild vegetables and mushrooms.

5 CONFLICTS OF INTEREST

There are no conflicts to declare.

6 ACKNOWLEDGEMENTS

We would like to express our gratitude to Dr. Susie Hardie for proofreading our manuscript.

7 REFERENCES

1. J. Anderson, N. Lévesque, F. Caron, P. Beckett and G. A. Spiers, *J. Environ. Radioact.*, 2022, **243**, 106797.

2. T. Dohi, Y. Ohmura, H. Kashiwadani, K. Fujiwara, Y. Sakamoto and K. Iijima, *J. Environ. Radioact.*, 2015, **146**, 125–133.

3. T. Dohi, K. Iijima, M. Machida, H. Suno, Y. Ohmura, K. Fujiwara, S. Kimura and F. Kanno, *PLoS ONE*, 2022, **17(7)**, e0271035.

4. T. Dohi, Y. Ohmura, K. Yoshimura, T. Sasaki, K. Fujiwara, S. Nakama and K. Iijima, *PLoS ONE*, 2021, **16(5)**, e0251828.

5. Hitachi High-Tech Corporation, *Electron microscopy techniques for structural cell biology*, **1(5)**, 2022 accessed, https://www.hitachi-hightech.com/jp/ja/knowledge/microscopes/sem-tem-stem/em-technology/

6. L. Fortuna, E. Baracchini, G. Adami and M. Tretiach, *J. Chem. Ecol.*, 2017, **43**, 1086–1096.

7. H. Jahns, *The lichens*, Academic press, New York, U.S., 1st edition, 1973.

8. R. Honegger, *Lichen Biology*, Cambridge University Press, Cambridge, U.K., 2nd edition, 2008 (1st published, 1996).

9. R. Honegger, *The New Phytologist*, 1986, **103**, 797–808.

10. R. Roth, R. Wagner and U. Goodenough, *Algal Research*, 2021, **60**, 102462.

CHAPTER 6 RADIONUCLIDES FOR HEALTH UK: IMPROVING UK ACCESS TO RADIONUCLIDES FOR MOLECULAR RADIOTHERAPY

J. D. Young[ab*], T. P. Tinsley[c], P. J. Blower[a] and J. K. Sosabowski[b]

[a]School of Biomedical Engineering and Imaging Sciences, King's College London, UK, [b]Centre for Cancer Biomarkers and Biotherapeutics, Barts Cancer Institute, Queen Mary University of London, UK, [c]The National Nuclear Laboratory, Sellafield, UK.

*jennifer.d.young@kcl.ac.uk

1 INTRODUCTION TO UK RADIONUCLIDE SUPPLY FOR MOLECULAR RADIOTHERAPY

Molecular radiotherapy (MRT) is the administration of a radiopharmaceutical for a therapeutic effect and is most commonly applied to late-stage metastatic cancer. The radioactive component of the radiopharmaceutical, the radionuclide, is delivered to the disease sites and accumulates there because of the radiopharmaceutical's targeting properties, either due to the radionuclide exploiting an endogenous pathway, such as ^{131}I-radioiodine accumulation in the thyroid, or because the radionuclide is attached to a targeting vector for a specific receptor (1). When the radionuclide decays the emissions provide a radioactive dose to the surrounding cells, which can cause radiation damage including single and double-strand DNA breaks, which in turn induces cell death, and provides the therapeutic effect.[1]

Radionuclides are an essential starting material for radiopharmaceuticals, and because of their short half-lives require a seamless supply chain from production, through radiopharmaceutical manufacture, to the point of patient care. Despite this, the UK is not self-sufficient in radionuclide production and instead imports all radionuclides for MRT and all radionuclides for diagnostics that have half-lives of over 2 hours. (Those with shorter half-lives are supplied by a UK network of cyclotrons).

The field of molecular radiotherapy (MRT) is currently experiencing unprecedented levels of interest and investment.[1] This has been driven by promising clinical results with beta-emitting radiopharmaceuticals and highly innovative and promising early clinical results with alpha-emitting radiopharmaceuticals. Beta-emitting radiopharmaceuticals include: (i) the recent approval (2017) of [^{177}Lu]Lu-DOTA-Tyr3-Octreotate[2,3] (Lutathera, AAA/Novartis) for treating somatostatin receptor-positive tumours and (ii) the Phase III randomised trial,[4] and FDA approval (March 2022),[5] and MHRA approval (August 2022)[6] of [^{177}Lu]Lu-PSMA-617 (Pluvicto, AAA/Novartis), a ligand targeted to the prostate specific membrane antigen (PSMA) for use in metastatic castrate-resistant prostate cancer (for which Endocyte raised $270 million in 2018 before being bought by Novartis for $2.1 billion). Among alpha-emitting radionuclides [^{225}Ac]Ac-PSMA-617 is the early front-runner for clinical application.[7] Hundreds of patients have been treated with this agent in Germany and South Africa under a compassionate use programme; remarkable anti-tumour efficacy has been seen in advanced-stage patients in retrospective studies and a clinical trial is now recruiting in South Africa and Australia (NCT04597411). However, the field of targeted alpha therapy (TAT) with ^{225}Ac and other alpha-emitting radionuclides such as ^{223}Ra, ^{211}At, ^{212}Pb/^{212}Bi, ^{227}Th and ^{148}Tb, and their future large-scale clinical application, are hampered by lack of supply. For instance, worldwide there are currently just four sources of the ^{229}Th required from which to extract ^{225}Ac. These are held in Germany, Canada, the USA and Russia, with a combined global annual production of ^{225}Ac of approximately 63 GBq (1.7 Ci).[8,9] This is enough for only around six thousand patient doses. Additional ^{229}Th stocks from US nuclear legacy material have the potential to increase this 40-fold,[10] and production of ^{225}Ac by other routes (including various accelerator-based routes) is an area of intense academic and industrial activity, as is the production of other therapeutic and diagnostic radionuclides.[8] As these stocks and development activities are happening overseas, it is difficult and expensive for UK researchers to access these materials and to date (Oct 2022), no trials of ^{225}Ac labelled radiopharmaceuticals have commenced in the UK.

Against this backdrop of high demand and poor supply (particularly in UK), the UK National Nuclear Laboratory (NNL), which has access to legacy nuclear material dating back to the 1960's, has opened discussions with UK-based nuclear medicine researchers, academics and industry on re-establishing the domestic supply of medical radionuclides in the UK. This would be in the form of existing stocks purified from UK nuclear legacy material, focusing on radionuclides that are not currently routinely available or which face future limitations on availability, including both beta and alpha particle emitters for use in molecular radiotherapy.

To facilitate these discussions with the academic community, and to advocate for radionuclide supply in the UK more widely, Radionuclides for Health UK has been established by Queen Mary University of London and King's College London researchers with expertise in nuclear medicine radiochemistry and the translation of radiopharmaceuticals (*https://www.bartscancer.london/radionuclides-for-health-uk/*). A key activity of Radionuclides for Health UK has been consolidating existing knowledge and capability in radionuclide production and building a national network of stakeholders from a range of disciplines to support a domestic supply of radionuclides. The initiative commenced with a Roundtable in November 2020 where three key production routes were identified. (i) Radionuclides from nuclear legacy material (ii) Reactor-produced radionuclides and (iii) Accelerator-produced radionuclides.

Between July 2021 and March 2022, three workshops were held, each focusing on one of the above production routes. The workshops were led by Radionuclides for Health UK with support from NNL (*www.nnl.co.uk*), CRUK RadNet City of London Radiation Research Unit (CRUK CoL RadNET) (*https://crukradnet.colcc.ac.uk*) and the Next Generation Molecular Imaging and Therapy with Radionuclides Programme (MITHRAS, an EPSRC-funded radiochemistry programme *https://www.mithrasprogramme.org/*). Each workshop included presentations from experts followed by panel discussions and was attended by representatives from academia, industry, government (Department for Business Energy and Industrial Strategy, Welsh Government Office for Science), funding bodies, and healthcare professionals. Feedback was collected from the workshop discussions and via a follow-up questionnaire. A report was produced on each topic summarising the content of the workshop and the views of the community. These workshops have galvanised the community around the need for domestic radionuclide supplies and highlighted routes to making this a reality. A summary of the workshops, the associated Radionuclides for Health UK white paper and the next steps for this initiative are provided below.

2 RADIONUCLIDES FROM LEGACY NUCLEAR MATERIAL

The Radionuclides for Health UK, National Nuclear Laboratory (NNL), CRUK CoL RadNet & MITHRAS workshop on Radionuclides from Legacy Nuclear Material was conducted online on 13th July 2021. Table 1 summarises the speakers and presentations at the workshop and the full workshop report is available on request (*https://www.bartscancer.london/radionuclides-for-health-uk/*).

The research community heard from NNL about the potential radionuclides that could be harvested from nuclear legacy material. All radionuclides produced by NNL would be useful to the research community. However, those available in large quantities, with feasible, high-purity, extraction processes are of most interest and should be prioritised. These include ^{212}Pb, ^{223}Ra and ^{90}Y. In addition, because of the limited worldwide supply, UK researchers would be interested in any quantity of ^{225}Ac that could be harvested, even if this was only on a scale suitable for pre-clinical research.

Overall, ^{212}Pb sourced from NNL provides the most opportunities for the UK research community. This is because it has attractive properties for molecular radiotherapy, including a half-life of 10.6 h which is well matched to the biological half-life of peptides and small molecules used to target cancer cells and because each of its two decay chains includes two β^- and one α emission. Crucially, it could be made available in quantities suitable for regular clinical use. Additionally, as ^{212}Pb-radiopharmaceuticals are less developed than ^{225}Ac-radiopharmaceuticals the UK could develop a cutting-edge and world-leading programme, due to this unique UK resource.

^{212}Pb is also the radionuclide with the most developed extraction process at NNL. ^{212}Pb decays to stable ^{208}Pb through beta and alpha emissions. Isotopically pure ^{228}Th is required to produce ^{212}Pb for use in molecular radiotherapy, and it is ^{228}Th that would be initially harvested by NNL from legacy nuclear material. NNL has established a research programme to extract ^{228}Th, ^{224}Ra and ^{212}Pb.[11] NNL has already developed a lab rig which can produce ^{212}Pb. Once the development has been optimised at this scale, processing could move onto a pilot plant module, for scale-up.

This pipeline assumes that the technology can produce ^{212}Pb of the purity and specification required for pre-clinical and clinical use, which is currently not proven. There are already UK academic centres interested in ^{212}Pb for chemistry and preclinical studies that would be able to start working with the material as soon as it is available. There are also several potential clinical studies that could be developed to build on the promising initial results from clinical trials on ^{212}Pb radiopharmaceuticals conducted in the USA.[12,13] Both academic and commercial interest has been shown in this radionuclide and the large quantity of ^{212}Pb potentially available from NNL could provide the UK with the resource and opportunity to become world-leading in the research and development of ^{212}Pb radiopharmaceuticals.

Table 1 *The speakers and talks at the Radionuclides for Health UK, National Nuclear Laboratory (NNL), CRUK CoL RadNet & MITHRAS workshop on Radionuclides from Legacy Nuclear Material.*

Speaker	Organisation	Talk Title
Jane Sosabowski	Queen Mary University of London, CoL RadNET, Radionuclides for Health UK	Opportunities for Improved Diagnosis and Treatments using Radiopharmaceuticals
Randy Perron	Canadian Nuclear Laboratory	Experience with Alpha Emitting Radionuclides
Howard Greenwood	UK National Nuclear Laboratory	Radionuclides with potential for use in Targeted Alpha Therapy
Richard Baldwin	Blueprint for growth contractor for UK National Nuclear Laboratory	UK Supply of Radionuclides from Harvesting
Phil Blower	King's College London, MITHRAS, CoL RadNET, Radionuclides for Health UK	UK Facilities – What we have and what we need

3 REACTOR-PRODUCED RADIONUCLIDES

The Radionuclides for Health UK, National Nuclear Laboratory (NNL), CRUK CoL RadNet & MITHRAS workshop on Reactor-Produced Radionuclides was conducted online on 23rd November 2021. Table 2 summarises the speakers and presentations at the workshop and the full workshop report is available on request (*https://www.bartscancer.london/radionuclides-for-health-uk/*). The workshop highlighted two key opportunities for the UK for research reactor-produced radionuclides:

(i) The research reactor at Institut Laue Langevin (ILL) Grenoble.

This reactor is co-funded, managed, and governed (one-third share) by the UK government through UKRI, although awareness of this resource to the nuclear medicine community has been historically poor. It forms one part of the PRISMAP EU network for radionuclide supply for research and produces a portfolio of radionuclides for research use including ^{177}Lu and ^{161}Tb. The workshop highlighted the opportunity to utilise this reactor as a resource and there was agreement across attendees that collaborations should be formed with ILL to provide both hands-on training and production of radionuclides that could then be shipped to the UK for processing, analysis and testing, further building up UK infrastructure and know-how.

(ii) The Welsh Government's ambitions for a new UK Research Reactor for radionuclide production – the Advanced Radioisotope Technology for Health Utility Reactor - ARTHUR.

Many of the reactors world-wide that produce medical radionuclides are likely to end their productive life over the next 10 years, and currently these account for a large proportion of the global medical radionuclide supply.[14] Therefore, there is a need for new reactors for radionuclide production, with currently two replacement projects underway, PALLAS in the Netherlands, and MYRRAH in Belgium. The Welsh Government has recognised this as an opportunity and proposes that a UK research reactor would improve the UK supply of medical radionuclides, and level up the UK to match Europe in terms of access to diagnostics and treatments using radionuclides. Welsh Government officials undertook an optioneering study (conducted by Rolls Royce PLC 2020) and considered accelerators, cyclotrons and six different reactor families. The study determined that the most suitable type of facility would be similar to the OPAL research reactor in Australia (Open Pool Australian Lightwater Research Reactor – which uses low-enriched uranium fuel). To determine the optimal site for such a reactor, thirty-four potential UK nuclear sites were narrowed to a top seven, and after more detailed investigation, the Trawsfynydd site in North Wales was identified as the best location. Ideally, both the reactor and processing plant would be at the same location in Wales (mimicking the Australian Nuclear Science and Technology Organisation (ANSTO) model located near Sydney).

The research community is supportive of the concept of a UK research reactor as this would provide stability of radionuclide supply and stimulation of research in the UK, which would lead ultimately to patient benefit. However, flexibility in the reactor design will be key. There must be flexibility to produce multiple radionuclides, enabling the UK to respond to changes in demand in clinical use and support research, as well as flexibility for the reactor to be used for other activities such as neutron scattering or nuclear materials testing. This will enable a business case to be built that includes both direct economic and indirect societal benefits for the UK. Such flexibility would be unlikely if it was a commercial reactor and so it is anticipated that government backing will be required, which is the model for the majority of research reactors globally. To be of most benefit to the UK a research reactor needs to be one part of the UK infrastructure. A pipeline from target material production, through irradiation, purification and radiopharmaceutical production is required as well as training schemes for skilled staff.

Table 2 *The speakers and talks at the Radionuclides for Health UK, National Nuclear Laboratory (NNL), CRUK CoL RadNet & MITHRAS workshop on Reactor-Produced Radionuclides.*

Speaker	Organisation	Talk Title
Allan Simpson	UK National Nuclear Laboratory	Reactor production of radionuclides
Sander de Groot	NRG	Reactor produced medical radionuclides and their supply
Luci Gleeson	Welsh Government	Advanced Radioisotope Technology for Health Utility Reactor (ARTHUR) Programme
Ulli Köster	Institut Laue Langevin (ILL) & Université Grenoble Alpes	Sustainability of Research Reactors Plus Emerging Radionuclides & PRISMAP
Eleanor Perkins	Advanced Nuclear Innovation Team, Department for Business, Energy and Industrial Strategy.	Advanced Nuclear Technologies Update
Lee Evitts	Nuclear Futures Institute, Bangor University	Incorporating medical radionuclide production into the nuclear industry

4 ACCELERATOR-PRODUCED RADIONUCLIDES

The Radionuclides for Health UK, National Nuclear Laboratory (NNL), CRUK CoL RadNet & MITHRAS workshop on Reactor produced Radionuclides was conducted online on 30th March 2022. Table 3 summarises the speakers and presentations at the workshop and the full workshop report is available on request from (*https://www.bartscancer.london/radionuclides-for-health-uk/*).

The UK has a network of low energy (<20 MeV) cyclotrons suitable for producing ^{18}F and ^{11}C radionuclides. In addition, some cyclotrons at research institutes, such as the Wales Research and Diagnostic Positron Emission Tomography Imaging Centre (PETIC), Cardiff University and King's College London are being used to produce other radionuclides. The UK has just one high-energy accelerator, which is located at the University of Birmingham, but this does not have on-site chemistry or GMP facilities and is used primarily for engineering applications.

Workshop discussions highlighted that for the UK to be able to compete with nuclear medicine research internationally, particularly to make the best use of the most advanced imaging technology (Total Body PET) and theranostic nuclear medicine, the UK would need to prioritise a sustainable and regular supply of accelerator-produced long- and short-lived radionuclides beyond fluorine-18. Both Cardiff University and King's College London have already started to respond to this need with programmes in place to develop ^{48}V and ^{89}Zr production at Cardiff University and ^{64}Cu, ^{124}I, ^{52}Mn, ^{203}Pb and ^{44}Sc production at King's College London.

In addition, access to target material is vital for the development of new accelerator production routes and UK nuclear legacy material should be considered as a novel source. There was support from the research community for a new high-energy accelerator facility. It would need to have the versatility to enable a range of radionuclides to be produced but also provide radionuclides regularly and consistently to best support laboratory research and clinical trials. It should be co-located with radiochemistry facilities and have excellent transport links to the whole of the UK.

Table 3 *The speakers and talks at the Radionuclides for Health UK, National Nuclear Laboratory (NNL), CRUK CoL RadNet & MITHRAS workshop on Accelerator-Produced Radionuclides.*

Speaker	Organisation	Talk Title
Veronika Rosecker	King's College London	Radionuclide production, selecting the best production method for the desired radionuclide
Allan Simpson	UK National Nuclear Laboratory	Accelerator Production of Radionuclides
Ferid Haddad	Arronax Nantes	Medical Radionuclide Production at ARRONAX Nantes. Overview of production facilities and plans to produce: ^{82}Sr, ^{211}At, ^{64}Cu, ^{44}Sc, ^{68}Ge, ^{67}Cu, ^{47}Sc
Renske Raaphorst	BV Cyclotron VU	Accelerator produced medical radionuclides at BV Cyclotron VU
David Parker	University of Birmingham	The MC40 Cyclotron Facility at the University of Birmingham
Karin Nielsen	King's College London, MITHRAS	Opportunities for cyclotron-produced radionuclides at KCL: Now and Future
Luiza Haberska	The Wales Research and Diagnostic Positron Emission Tomography Imaging Centre (PETIC), Cardiff University	Cyclotron Production of Radionuclides at PETIC: ^{18}F, ^{13}N, ^{48}V, ^{89}Zr.
Seán Collins	Nuclear Metrology Group, National Physical Laboratory	Metrology for Medical Radionuclides

5 WHITE PAPER - RADIONUCLIDE SUPPLY IN THE UK: A PATH TO CANCER BREAKTHROUGH

Radionuclides for Health UK have taken the learnings from the workshops plus discussions with stakeholders and produced a White Paper - *Radionuclide Supply in the UK: A Path to Cancer Breakthrough* - which was launched in October 2022. A full version of the white paper can be found at *https://www.bartscancer.london/radionuclides-for-health-uk/*.

This white paper summarised the opportunities afforded by molecular radiotherapy treatment, not only to improve patient outcomes but also to provide the UK with the opportunity to establish itself as a world leader in expanding this innovative and highly personalised medicine. After describing the challenges and opportunities the white paper goes on to outline a five-point plan to help accelerate this ambition outlined below:

1. Invest in nuclear infrastructure for medical radionuclides in the UK.
2. Develop enhanced nuclear processing capabilities.
3. Create a new research-friendly regulatory framework for nuclear medicines.
4. Create a centralised, nationally accessible, appropriately resourced radiopharmacy dedicated to research and trials.
5. Close the UK skills gap through expanded training opportunities and regulatory reform allowing more people to train to work in the sector.

Radionuclides for Health UK has initiated activities to support these five areas and will continue to advocate and take action to realise these goals.

6 CONCLUDING REMARKS

Molecular radiotherapy has the opportunity to transform cancer treatment and the UK could take advantage of this by investing in infrastructure for radionuclide production and processing. Radionuclides can be produced from legacy nuclear material, research reactors and accelerators, and all three production methods are needed to provide a full portfolio of radionuclides for medical applications and research use. Three workshops have been held to gain input from experts and highlight interests and opportunities for the research community. This has led to the development of a White Paper which summarises the investment and actions needed for the UK to secure radionuclide supply.

7 REFERENCES

1. L. Bodei, K. Herrmann, H. Schöder, A.M. Scott, J.S. Lewis, *Nature Reviews Clinical Oncology. Nature Research*; 2022, **19**, p. 534–50.

2. National Institute for Health and Care Excellence. NICE Guidance. Lutetium (177Lu) oxodotreotide for treating unresectable or metastatic neuroendocrine tumours. Technology appraisal guidance [Internet]. 2018. Available from: https://www.nice.org.uk/guidance/ta539

3. European Medicines Agency. Assessment report. EMA/506460/2017 Committee for Medicinal Products for Human Use (CHMP) Lutathera. International non-proprietary name: lutetium (177lu) oxodotreotide [Internet]. Vol. 44. 2017. Available from: www.ema.europa.eu/contact

4. K. Rahbar, L. Bodei, M.J. Morris, *Journal of Nuclear Medicine*. 2019; **60**, 1504.

5. U.S. Food & Drug Administration. FDA approves Pluvicto for metastatic castration-resistant prostate cancer [Internet]. 2022 [cited 2022 Oct 24]. Available from: https://www.fda.gov/drugs/resources-information-approved-drugs/fda-approves-pluvicto-metastatic-castration-resistant-prostate-cancer#:~:text=On%20March%2023%2C%202022%2C%20the,metastatic%20castration%2Dresistant%20prostate%20cancer%20(

6. Novartis. MHRA grants marketing authorisation for Novartis' radioligand therapy Pluvicto®▼ (lutetium (177Lu) vipivotide tetraxetan) in advanced prostate cancer in Great Britain [Internet]. 2022 [cited 2022 Oct 24]. Available from: https://www.novartis.com/uk-en/news/media-releases/mhra-grants-marketing-authorisation-novartis-radioligand-therapy-pluvictov-lutetium-177lu-vipivotide-tetraxetan-advanced-prostate-cancer-great-britain

7. C. Kratochwil, F. Bruchertseifer, H. Rathke, M. Hohenfellner, F.L. Giesel, U. Haberkorn, et al. *Journal of Nuclear Medicine*. 2018, **59**, 795.

8. V. Radchenko, A. Morgenstern, A. Jalilian, C. Ramogida, C.S. Cutler, C. Duchemin, et al. *Journal of Nuclear Medicine*, 2021, **62,** 1495.

9. A. Morgenstern, C. Apostolidis, F. Bruchertseifer, *Seminars in Nuclear Medicine*. 2020, **50**, 119.

10. TerraPower A Nuclear Innovation Company. Developing Advanced Cancer Treatments Using Actinium-225 [Internet]. 2021 [cited 2021 Nov 9]. Available from: https://www.terrapower.com/medical-isotopes-actinium225/

11. UK National Nuclear Laboratory. NNL makes breakthrough to address the global shortage of key medical radioisotope for fighting cancer [Internet]. nnl.co.uk. 2022 [cited 2022 Oct 24]. Available from: https://www.nnl.co.uk/2022/05/nnl-makes-breakthrough-to-address-the-global-shortage-of-key-medical-radioisotope-for-fighting-cancer/

12. E.S. Delpassand, I. Tworowska, R. Esfandiari, J. Torgue, J. Hurt, A. Shafie, et al. *Journal of Nuclear Medicine*, 2022, **63,** 1326.

13. E. Delpassand, R. Esfandiari, I. Tworowska, J. Torgue, J.D. Hurt, R. Nunez, *Journal of Clinical Oncology*, 2022, **40 no.16 sup**, 4128.

14. CEA, NCBJ, PALLAS, NRG, SCK•CEN, TUM, et al. European Research Reactor Position Paper for DGE Energy-2018 report [Internet]. 2018 [cited 2022 Oct 24]. Available from: https://www.frm2.tum.de/fileadmin/w00bnv/www/Neutronenquelle/Bestrahlungsanlagen/European_Research_Reactor_Position_Paper_for_DGE_Energy__2018_report_20180801.pdf

CHAPTER 7 DETERMINATION OF BLUE CARBON SEQUESTRATION RATES IN THE UK NORTH SEA USING LEAD-210 AND COMPLEMENTARY FINGERPRINTING TOOLS

Franck Dal Molin[a*], Paul Blowers[a], Daniel Brady[a], Nathan Brown[a], Stephanie Cogan[a], Terri Davis[a], Peter Hamstead[a], Mariusz Huk[a], Hannah Limbach[a] and Paul Smedley[a]

[a]Centre for Environment, Fisheries and Aquaculture Science, Lowestoft, NR33 0HT, UK

*franck.dal-molin@cefas.gov.uk

1 INTRODUCTION

The study of the historical patterns of environmental changes, variations and pressures caused by human activities in sediment cores has several applications.[1,2] They provide insights to local and global time trends of past and present use because sediments in relatively undisturbed aquatic systems can be used as temporal integrators. This process of assigning the age to layers of marine core sediment can be achieved by determining the fraction of ^{210}Pb present from the decay of radon gas in the atmosphere (also called unsupported fraction) and by correlating the subsequent deposits in sedimentary substrates with any substances of interest.[3,4] While the multidisciplinary features of ^{210}Pb-based modelling approaches are well known across the scientific community, their application can be very challenging depending mainly on the ^{210}Pb concentration levels present in the sediment and the levels of precision and accuracy required.[5,6] The data interpretation is also a difficult exercise as many biogeochemical and human factors must be considered when analysing resulting ^{210}Pb depth profiles.[7] Indeed, natural bioturbation, resuspension as well as trawling activities can complicate the straightforward determination of sediment accumulation rates from ^{210}Pb depth profiles. Furthermore, other sources of ^{210}Pb present in the sediment, from the natural decay of *in situ* ^{226}Ra and, more problematically, from non-nuclear industrial activities discharging effluents enriched in ^{226}Ra and ^{210}Pb can make the determination of the unsupported ^{210}Pb fraction very difficult. To resolve this challenge, characteristic discharges from major nuclear or industrial activities are generally used as fingerprints to validate sedimentation rate information obtained from ^{210}Pb-based modelling. In the majority of research studies undertaken in the Northern hemisphere, the ^{137}Cs fingerprints present from legacy nuclear weapon testing in the late 1950s and early 1960s or from the Chernobyl accident in 1986 have been extensively used to validate ^{210}Pb core profiles.[8,9]

These tools are currently of interest to marine scientists who are trying to refine knowledge of climate change impact on aquatic ecosystems and, more precisely, improve understanding of how fast our coastal and marine vegetative systems, also known as BC ecosystems, store and sequester CO_2 in the long term. BC is more generally a term used to refer to organic carbon that is initially captured by photosynthesis in algae, seagrass, mangroves, and other plants growing in salt marshes and coastal wetlands. Understanding of the capacity of marine sediments to subsequently store and sequester BC is key to assessing the potential for management of human activities to protect these stores and maximise sequestration potential as nature-based solutions to climate change. Presently, the evidence base relevant to carbon sequestration remains relatively poor, particularly in the offshore UK marine environment.[10]

Several sea surveys were conducted between 2019 and 2021 to determine BC sequestration rates in three different regions of the UK North Sea: (A) the Outer Silver Pit area, (B) the Berwick to St. Mary's Marine Conservation Zone (MCZ) and (C) 50 metres away from a decommissioned Oil & Gas (O&G) platform located in the northern part of UK North Sea between Scotland and Norway. The Outer Silver Pit area is an offshore submarine valley in the bed of the North Sea, located between the Dogger bank and the ridge dividing the northern from the southern North Sea basins, and known to be intensively affected by trawling. The Berwick to St. Mary's MCZ is a near shore site located along the Northumberland coast in the north-east part of England and assumed to be only affected by natural coastal dynamic processes. The third selected sampling location in this study is offshore and was assumed to be affected mainly by submarine man-made structures and legacy waste discharges from the production of O&G.[11]

These surveys at sea were an opportunity to (i) trial different sample preparative and analytical approaches for measuring low levels of ^{210}Pb in marine sediment samples, (ii) assess the impact of natural and / or anthropogenic disturbances associated with the selected sites, (iii) test one of the most common ^{210}Pb-based modelling approaches and (iv) investigate the best fingerprinting tools available for validating BC sequestration rates in these specific marine regions, affected by different anthropogenic pressures and presenting different hydrodynamic and sedimentological regimes.

2 MATERIAL AND METHODS

2.1 Sample collection and preparative methods

Sediment cores were collected using a multicorer at 3 different locations in the North Sea (see Figure 1). The multicorer used in this study was designed by the Royal Netherlands Institute for Sea Research (NIOZ) and allowed the simultaneous sampling of up to 12 tubes of 10 cm diameter in a relatively undisturbed manner. The maximum sediment depth recoverable with this coring equipment is generally around 25 cm, but this was only achieved at location B (see Figure 1), where a high concentration of fine sediments (mud) could be observed. In coarser sediments (e.g. sands, gravels), present at sampling locations A and C, it was only possible to core down to 20 and 13 cm, respectively.

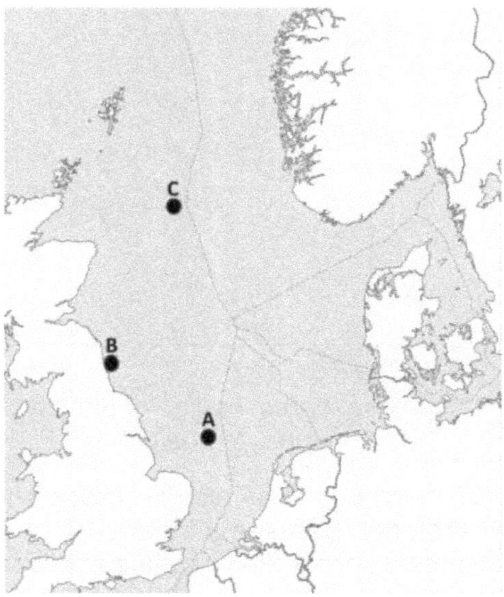

Figure 1 *Map of the North Sea showing the three sampling locations selected in this study.*

The sediment cores collected from locations A and C were then sliced every centimetre (cm) and the core collected from location B was sliced every 1 or 2 cm down to 25 cm, with an additional 7 cm slice between 25 and 32 cm. All core slices were then freeze dried for up to 3 days using a device manufactured by Frozen In Time (UK), allowing multiple samples to be processed simultaneously. All core slices collected from location A were sieved down to 1 mm and slices from sampling locations B and C to 250 μm. The full physical preparation of sediment prior to analysis is summarised and schematically represented in Figure 2.

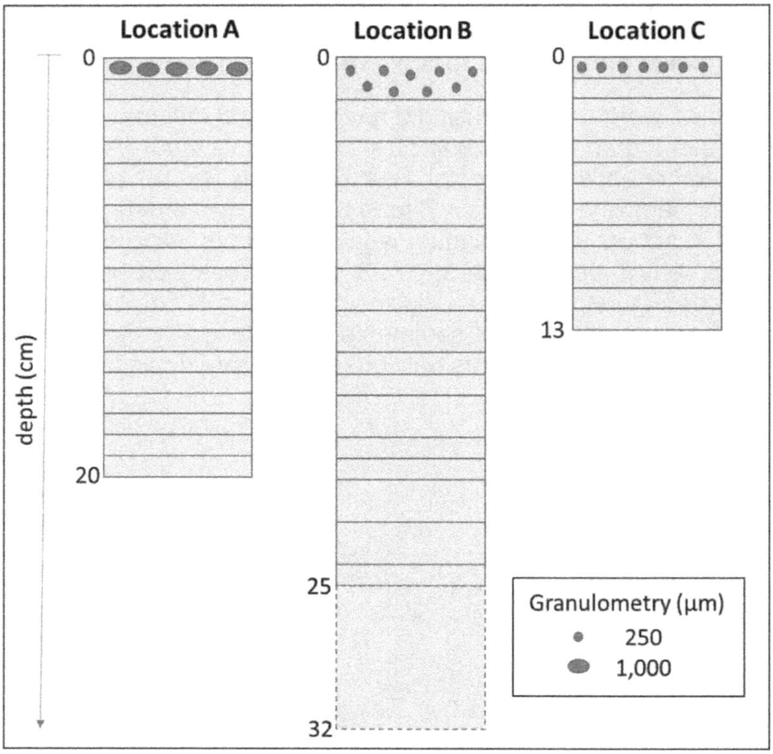

Figure 2 *Schematic representation of the physical sample preparation prior to analysis.*

The particle size distribution was also assessed and found to be relatively constant across each core. Between 27 and 34 g of dried material was then transferred and compressed in a plastic container prior to gamma analysis (see Figure 3). Samples requiring additional ^{226}Ra analysis were also sealed using silicon sealant to minimise radon gas losses and ensure secular equilibrium between ^{226}Ra and its decay product, ^{214}Pb after 21 days (see more details in Section 2.2).

Figure 3 *Picture of a sealed container of sediment for gamma analysis.*

2.2 Radioanalytical methods

Total ^{210}Pb and ^{137}Cs activity concentrations were determined in most of the samples by gamma spectrometry, using Broad Energy Germanium detectors (ORTEC, USA) and their respective characteristic emission peaks of 46.5 and 662 keV. A mixed radionuclide standard solution in 4M HCl (NPL, UK) and the certified reference material (CRM) IAEA-465 (Baltic Sea sediment) were also used to perform the energy and efficiency calibrations of each detector used in this study.

To validate our results internally, total ^{210}Pb was also determined in many slices of the core collected from sampling location A by alpha spectrometry via its decay product, ^{210}Po. In more detail, sample aliquots of approximately 1.5 g were spiked using a known amount of ^{208}Po tracer (NPL, UK) and digested in 100 ml of *Aqua Regia* on a hotplate overnight. The dried samples were taken up in 50 ml of 6 M HCl solution, refluxed for 1 h, reduced to 25 ml and filtered. The samples were reduced further down to approximately 5 ml and taken up to a final volume of 35 ml using 30 ml of 0.5 M HCl. A few milligrams of ascorbic acid were added to the solution to reduce iron III interfering with polonium during the electrodeposition process onto silver discs. The discs were then rinsed with deionised water, industrial methylated spirit, dried and counted using Passivated Implanted Planar Silicon (PIPS) detectors (ORTEC, USA). After approximately 5 days of counting, ^{210}Po could then be quantified by isotope dilution and was assumed to be in secular equilibrium with ^{210}Pb at sampling time. The results were also validated using the CRM IAEA-385 (Irish Sea sediment) and an internal quality control material.

The activity concentration of supported ^{210}Pb present in sediment samples preliminarily prepared in sealed containers was indirectly determined by gamma spectrometry using the ^{214}Pb peak at 352 keV (assumed in secular equilibrium with ^{226}Ra). This approach was used to demonstrate the naturally occurring ^{226}Ra concentration in the two cores collected from the sampling locations A and B remained constant. ^{226}Ra was also determined in each slice of the core collected from sampling location C using the same analytical method to investigate the presence of anthropogenic ^{226}Ra (an additional source of supported ^{210}Pb) from legacy discharges of produced water. The full radioanalytical programme is presented in Figure 4.

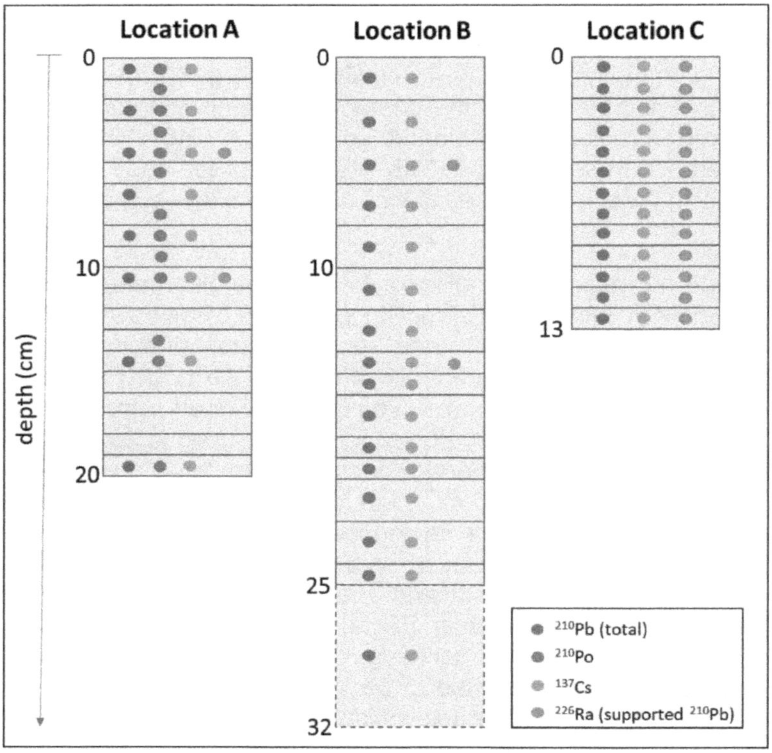

Figure 4 *Schematic representation of the radioanalytical methodologies.*

2.3 Modelling and fingerprinting approaches

The proportion of ^{210}Pb already present in the sediment and from any other anthropogenic activities (termed supported ^{210}Pb, see Figure 5) does not decay at the same rate as the unsupported ^{210}Pb fraction from atmospheric deposition used in ^{210}Pb-based modelling.

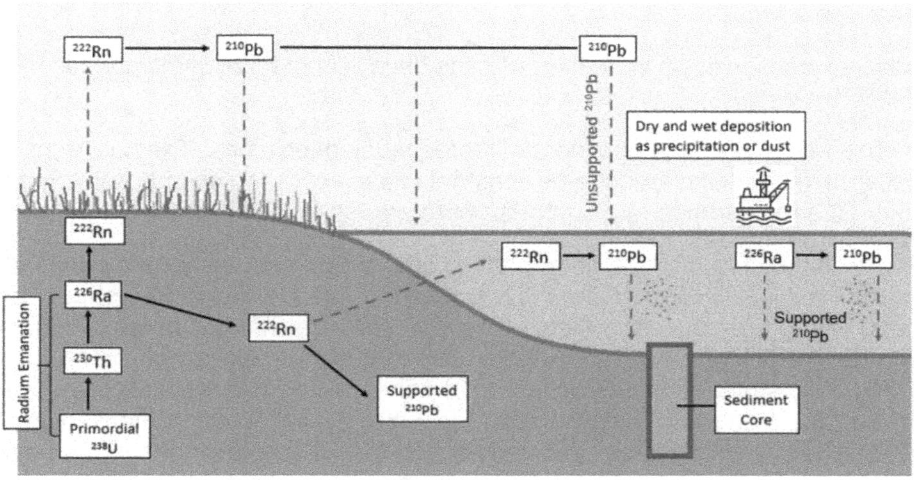

Figure 5 *^{210}Pb based modelling principles.*

Measuring ^{226}Ra in addition to ^{210}Pb enables the subtraction of supported ^{210}Pb from the total ^{210}Pb prior to estimating sedimentation rates and dating sediment layers. With a half-life ($t_{1/2}$) of 22.23 years, ^{210}Pb dating is limited to relatively recent sediment ages (up to 150 years). It can therefore be challenging to use this technique in heavily disturbed marine environments with a sedimentation rate lower than 0.1 cm y^{-1}.[12] There are several ^{210}Pb-based models that can be applied and help interpret ^{210}Pb core profiles. The Constant Initial Concentration (CIC)[13] and Constant Rate of Supply (CRS)[14] models are the most widely used to estimate sedimentation or accumulation rates, but both are associated with numerous assumptions that can sometimes limit their appropriateness in more complex and disturbed aquatic ecosystems.[7] In addition, the mass accumulation rate, expressed in mg cm^{-2} y^{-1}, has also been estimated in many previous studies to account for compaction effect.[15]

The CRS model assumes that the rate of supply of unsupported ^{210}Pb to sediments from the atmosphere at a core location remains constant, irrespective of variations in the mass accumulation rate (accumulation rate for dry mass of sediments). This model is more reliable than the CIC model, with the assumption that the initial ^{210}Pb concentrations were constant for each layer at the time of its formation is highly unlikely to be realistic in most deposition environments.[16]

When using the CIC model, the sedimentation rate S or, in this study, the sequestration rate at a specific location can be estimated by plotting the core depth against the natural logarithm of concentrations of unsupported ^{210}Pb and attempting to extract a correlation with a minimum of 5 data points across the core. The resulting slope from each correlation can then be directly related to the sequestration rate S and can be expressed as:

$$S = \text{slope} \times \lambda_{210Pb} \quad [1]$$

with S expressed in cm y^{-1} and λ_{210Pb} is the decay constant of unsupported ^{210}Pb of 0.031 y^{-1}.

The CIC model was applied at all three studied locations. The bulk density was also verified and found to be constant in each core (no compaction effect). Thus, all sequestration rates reported here are expressed in cm y^{-1}.

The attribution of ^{137}Cs fingerprint to late 1950s and early 1960s nuclear weapon testing or to the Chernobyl accident in the UK North Sea should be made with great caution in view of the potential presence of other important sources, particularly from the legacy discharges from Sellafield in the mid-1970s.[17] This other source has already been detected in cores collected near UK shores, along Greenland and in the Baltic Sea.[18] Routine monitoring of ^{137}Cs in seawater samples collected around the UK shows that a ^{137}Cs peak observed in the Irish Sea near Sellafield in 1974-75 could also be detected residually in 1978-79 in both the southern and northern parts of the UK North Sea, suggesting a transient time of 3-5 years around the Scottish coast (see Figure 6).[19]

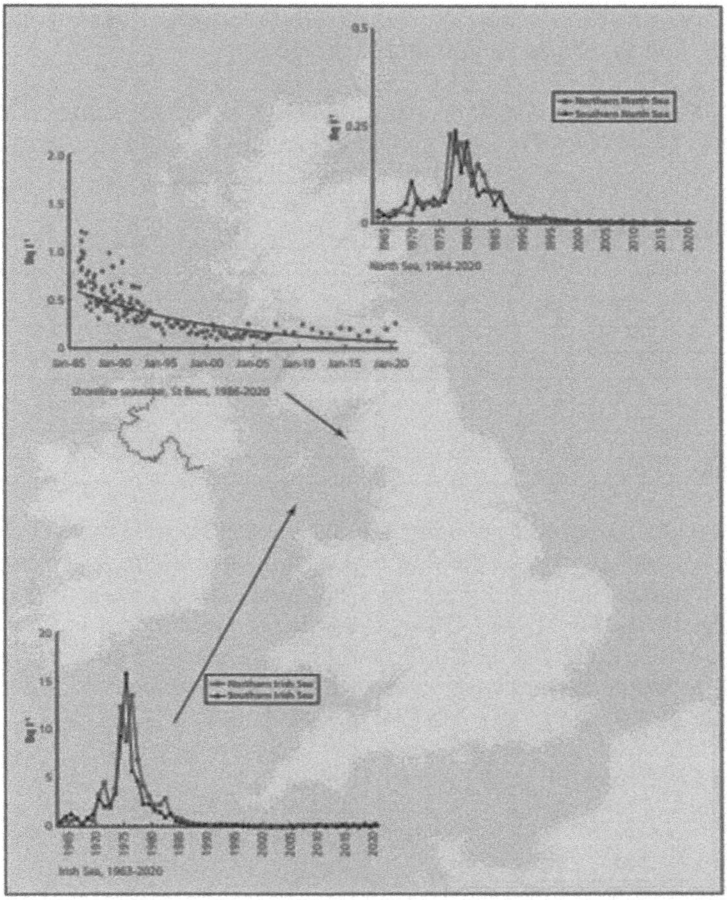

Figure 6 *^{137}Cs concentration in the Irish Sea, North Sea and shoreline. Reproduced from ref. 19 under the terms of the Open Government License, https://www.nationalarchives.gov.uk/doc/open-government-licence/version/3/.*

In addition to the Sellafield ^{137}Cs fingerprint, the potential presence of anthropogenic ^{226}Ra at the sampling location C was also investigated as a potential alternative fingerprinting option to validate the sedimentation rate obtained from ^{210}Pb CIC modelling. Selected marine sediments collected from a separate study in the direct vicinity of a wastewater discharge site were found to contain elevated levels of ^{210}Pb, ^{226}Ra and its chemical analogues such as barium and strontium.[20]

3 RESULTS AND DISCUSSION

3.1 ^{210}Pb measurement: comparison of methods and validation

The concentrations of total ^{210}Pb in several slices of the sediment core collected from the Outer Silver Pit area (sampling location A) were determined by gamma spectrometry and indirectly by alpha spectrometry, assuming secular equilibrium. The results are summarised in Table 1.

Table 1 *^{210}Pb and ^{210}Po activity concentrations in sediment samples collected from the Outer Silver Pit area (sampling location A) and comparison with IAEA-385 certified reference value.*

Depth (cm)	^{210}Po (Bq kg^{-1})	1σ (Bq kg^{-1})	^{210}Pb (Bq kg^{-1})	1σ (Bq kg^{-1})
0 - 1	19.6	0.8	19.4	1.7
2 - 3	15.8	0.7	19.1	1.5
4 - 5	18.4	0.8	19.8	1.8
7 - 8	16.6	0.9	18.2	1.9
9 - 10	13.8	0.6	18.0	1.6
13 - 14	12.9	0.4	16.9	1.5
14 - 15	12.4	0.4	14.3	1.4
19 - 20	11.6	0.3	15.6	1.6
IAEA - 385	29	1	32	4

Many results disagreed between the two methods, with most results from gamma spectrometry analysis being higher than results obtained from alpha spectrometry analysis (see Figure 7). This positive bias was also observed when analysing the CRM IAEA-385, presenting a decay corrected target ^{210}Pb concentration of 29 ± 1 Bq/kg. Due to the higher uncertainties associated with gamma spectrometry analysis, ^{210}Po data were used to determine a sedimentation rate using the ^{210}Pb CIC model (Equation [1]).

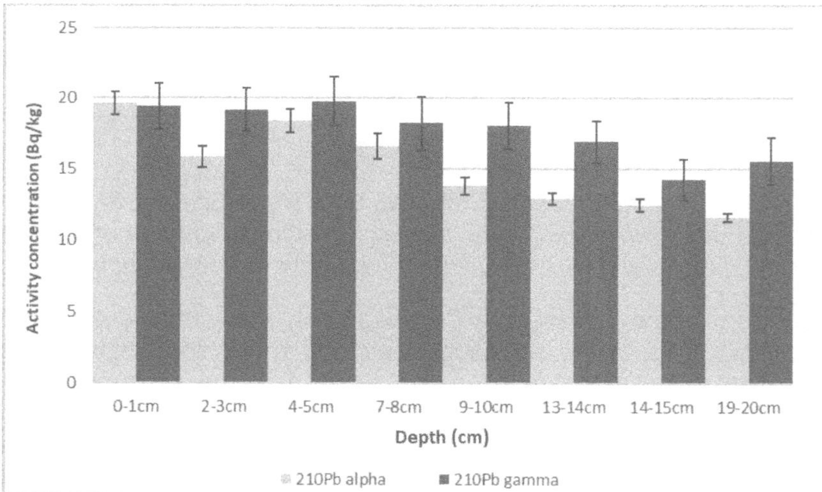

Figure 7 *Method comparison for measuring ^{210}Pb in sediment core slices collected from the Outer Silver Pit area (sampling location A).*

3.2 Blue Carbon sequestration rates

All results from the surveys undertaken at sampling locations A, B and C are summarised in Figure 8. The ^{210}Po (total ^{210}Pb) profile obtained at sampling location A (Outer Silver Pit area) was very disturbed within the top 9 cm of the core and therefore could not be exploited. ^{226}Ra concentration was confirmed to be constant across the core and was found to be 8.9 Bq/kg on average. This enabled the calculation of unsupported ^{210}Pb in several slices of the core segment between 9 and 20 cm. The sedimentation rate or BC sequestration rate could then be estimated at 0.51 cm y^{-1}, using 5 data points with a R^2 of 0.99 and a slope of 16.46. Unfortunately, this result could not be validated using a ^{137}Cs fingerprint, as most values were close to its limit of detection by gamma spectrometry (1 Bq/kg). The small increase of ^{137}Cs observed at 19-20 cm could not be fully attributed to the 1978-79 Sellafield-derived peak and suggested the need to repeat sampling, core deeper than 20 cm in this region of the UK North Sea and investigate other fingerprinting options. Overall, the concentration levels observed in this study were very low. Alpha spectrometry was used as an alternative to determine total ^{210}Pb in an indirect manner, but this approach requires labour intensive sample preparative steps. Two slices were selected and sieved down to 250 µm to quantify the impact of finer granulometry on the absolute total ^{210}Pb activity concentrations and explore further the possibility for measuring ^{210}Pb directly by gamma spectrometry and lowering uncertainties. The increase was found to be more than 40% and considered sufficient for using this more direct and non-destructive radiometric approach at the other two sampling locations.

Table 2 *Results from all surveys undertaken at sampling locations A, B and C.*

Sampling location	Depth	^{210}Pb total[a]	^{226}Ra	^{137}Cs	^{210}Pb unsupported	Slope[b]	R^2	S (^{210}Pb CIC)	S (fingerprint[c])
	(cm)	(Bg/kg)	(Bq/kg)	(Bq/kg)	(Bg/kg)			(cm/y)	(cm/y)
A	0-1	19.6 ± 0.8		1.0 ± 0.2					
	1-2	14.2 ± 0.6		n.d.					
	2-3	15.8 ± 0.7		0.9 ± 0.2					
	3-4	20.8 ± 0.8		n.d.					
	4-5	18.4 ± 0.8	9.5 ± 0.7	1.0 ± 0.2					
	5-6	19.7 ± 0.8		n.d.					
	6-7	n.d.		0.9 ± 0.2					
	7-8	16.6 ± 0.9		n.d.					
	8-9	13.6 ± 0.6		0.8 ± 0.2					
	9-10	13.8 ± 0.6		n.d.	4.9 ± 1.1	16.46	0.99	0.51	0.47[d]
	10-11	13.6 ± 0.5	8.3 ± 0.7	0.8 ± 0.2	4.7 ± 1.1				
	14-15	12.9 ± 0.4		n.d.	4.0 ± 1.0				
	15-16	12.4 ± 0.4		1.0 ± 0.2	3.6 ± 1.0				
	19-20	11.6 ± 0.4		1.4 ± 0.2	2.7 ± 0.9				
B	0-2	47.1 ± 3.4		2.3 ± 0.2					
	2-4	54.6 ± 3.8		4.0 ± 0.3					
	4-6	46.7 ± 3.2	15.4 ± 0.9	3.6 ± 0.3					
	6-8	36.8 ± 2.6		2.8 ± 0.2					
	8-10	38.1 ± 2.7		3.0 ± 0.3					
	10-12	42.5 ± 3.2		4.4 ± 0.3	27.3 ± 3.3	10.51	0.99	0.33	0.35

	12-14	34.2 ± 2.7		5.6 ± 0.4	19 ± 2.9				
	14-15	28.6 ± 2.9	15.0 ± 0.8	6.6 ± 0.4	n.i.				
	15-16	31.9 ± 2.6		7.2 ± 0.4	16.7 ± 2.7				
	16-18	25.6 ± 2.1		5.0 ± 0.3	n.i.				
	18-19	27.9 ± 2.6		3.2 ± 0.3	12.7 ± 2.7				
	19-20	21.1 ± 2.2		2.0 ± 0.2	n.i.				
	20-22	19.9 ± 2.9		2.3 ± 0.3	n.i.				
	22-24	17.2 ± 2.0		1.5 ± 0.2	n.i.				
	24-25	22.1 ± 2.6		1.4 ± 0.2	n.i.				
	25-32	20.1 ± 2.7		1.3 ± 0.2	4.9 ± 2.8				
C	0-1	63.9 ± 4.8	159.1 ± 6.2	0.9 ± 0.2					
	1-2	23.8 ± 2.6	42.8 ± 1.8	3.2 ± 0.3					
	2-3	17.5 ± 3.1	25.8 ± 1.1	2.9 ± 0.3					
	3-4	10.5 ± 2.3	26.7 ± 1.2	3.3 ± 0.3					
	4-5	14.2 ± 2.2	25.8 ± 1.2	2.8 ± 0.3					
	5-6	16.9 ± 3.3	23.3 ± 1.1	2.5 ± 0.3					
	6-7	13.6 ± 2.6	24.7 ± 1.2	2.9 ± 0.3					
	7-8	23.1 ± 3.8	27.9 ± 1.3	2.8 ± 0.3					
	8-9	20.8 ± 2.9	19.7 ± 1.7	2.1 ± 0.2	10.1 ± 3.0	3.41	0.61	0.11	0.13[e]

[a]Determined via ^{210}Po by alpha spectrometry at sampling location A
[b]Corresponding to S/λ^{210}Pb with S: sedimentation rate; λ^{210}Pb: ^{210}Pb decay constant = 0.031 y^{-1}
[c]Use of Sellafield-derived ^{137}Cs at sampling locations A and B and ^{226}Ra from legacy discharges of produced water at sampling location C
[d]Not fully validated
[e]Determined by assuming no legacy discharges between end of decommissioning process in 2007 and 2021 (sampling year)
n.d. Not determined
n.i. Data point not included in ^{210}Pb-based CIC model

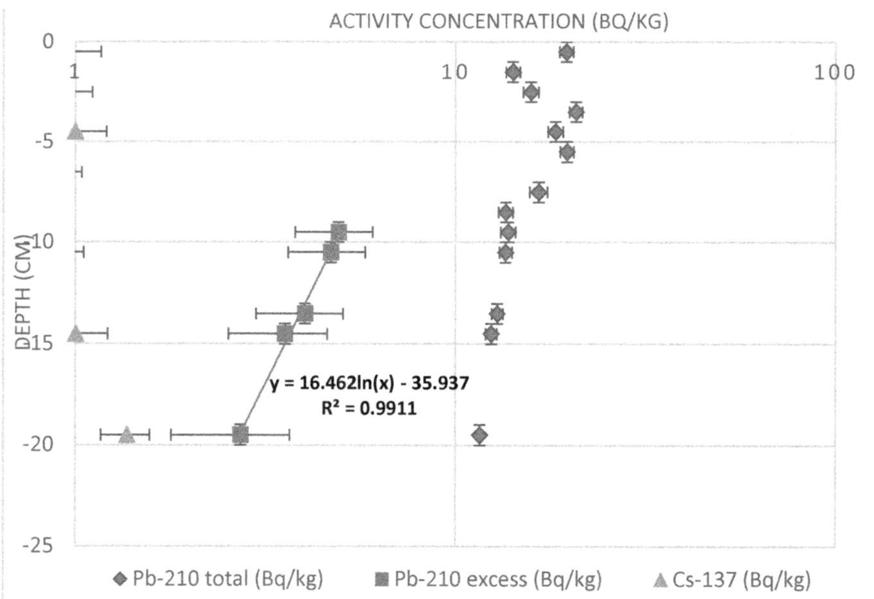

Figure 8 *Total ^{210}Pb, unsupported ^{210}Pb, ^{137}Cs profiles and CIC modelling for the core collected from location A (Outer Silver Pit area).*

As shown in Figure 9, disturbances were also observed in the top 10 cm of the core sampled near shore within the Berwick - St. Mary's MCZ (sampling location B). A clear correlation could though be extrapolated within the core segment between 10 and 32 cm. The sedimentation rate was estimated to be 0.33 cm y^{-1}, using 5 data points, with a R^2 of 0.99 and a slope of 10.505. A clear ^{137}Cs peak was also observed around 14-15 cm. A sedimentation rate of 0.35 cm y^{-1} was determined by assuming it was derived entirely from the transient Sellafield 1978-79 peak and this hypothesis was therefore verified and enabled to confirm the sedimentation rate result estimated from ^{210}Pb CIC modelling.

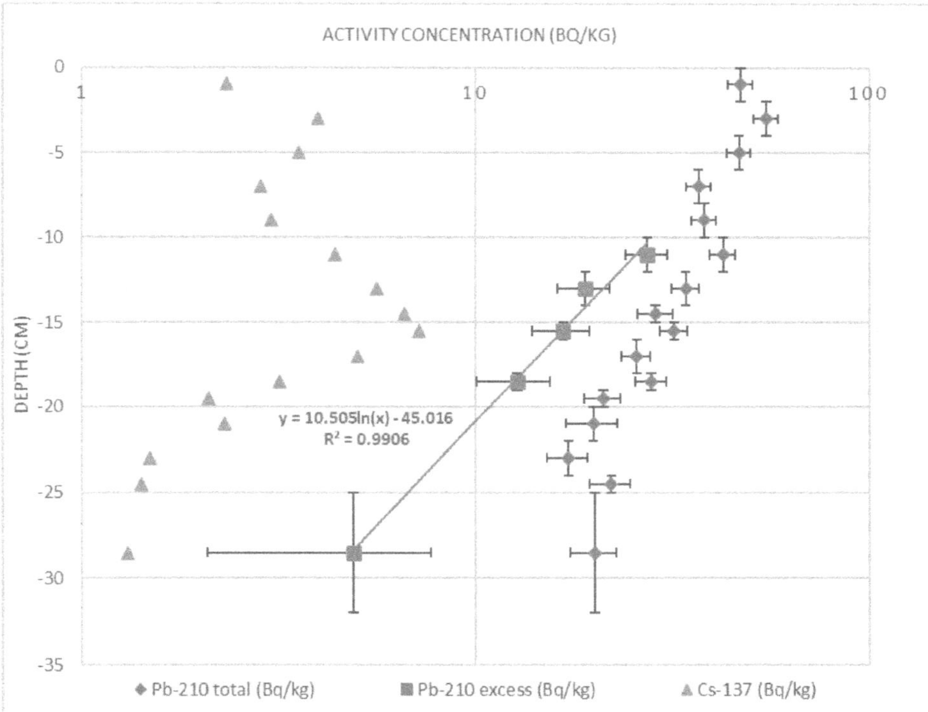

Figure 9 *Total ^{210}Pb, unsupported ^{210}Pb, ^{137}Cs profiles and CIC modelling for the core collected from location B (Berwick - St. Mary's MCZ).*

As can be seen in Figure 10, disturbance was observed within the top 8 cm of the core but no obvious ^{137}Cs peak could be distinguished and therefore used as a fingerprint tool to validate the results from ^{210}Pb CIC modelling. Elevated ^{226}Ra concentrations and, to a lesser extent ^{210}Pb, were also noted in the top 10 cm, with the highest concentrations detected within in the top 1-2 cm. This was a clear signature of legacy discharges of produced water from the O&G platform in the direct vicinity of the sampling location C, now fully decommissioned since 2007. The concentration of ^{226}Ra seemed to stabilise below 10 cm, implying the naturally supported ^{210}Pb fraction to be just above 10 Bq/kg (10.7 Bq/kg on average from 3 measurements made between 10 and 13 cm). This site-specific fingerprinting approach enabled to confirm that the sedimentation rate result was less than 0.13 cm y^{-1}, and within the same range as the result obtained from ^{210}Pb-based CIC modelling (0.11 cm y^{-1}), with moderate certainty (R^2 of 0.61) and using 5 data points between 8 and 13 cm.

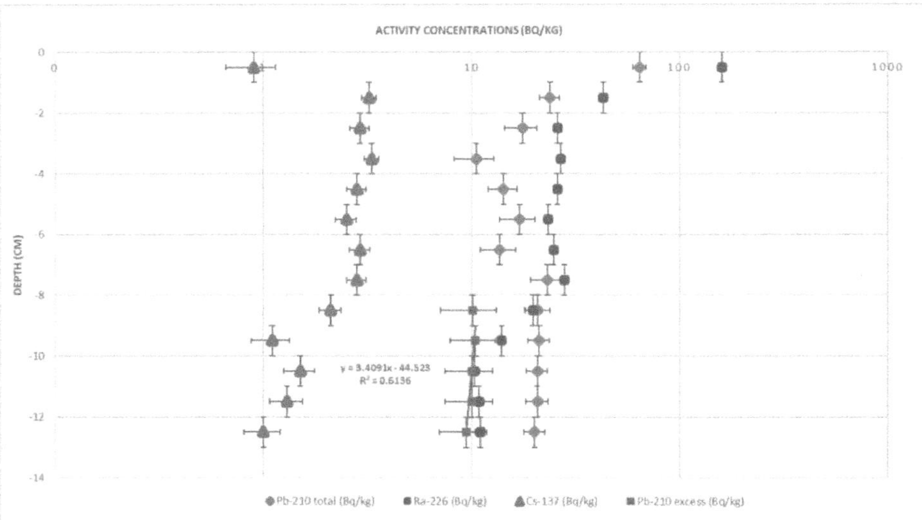

Figure 10 *Total ^{210}Pb, unsupported ^{210}Pb, ^{137}Cs profiles and CIC modelling for the core collected from location C (offshore oil and gas platform).*

4 CONCLUSIONS

Blue Carbon (BC) habitats are one of the largest natural carbon sinks on Earth that need to be protected or restored to help mitigate the risk from climate change.[10] The ^{210}Pb CIC modelling approach enables the estimation of BC sequestration rates in the UK North Sea seabed, but many considerations must be taken prior to organising sampling surveys at sea. Indeed, BC sequestration rates were estimated from sediment cores collected from three different regions of the UK North Sea between 2021 and 2022 and were found to be ranging between approximately 0.1 and 0.5 cm y^{-1}. This variability requires the implementation of different sampling, preparative and analytical strategies to maximise chances for obtaining a good correlation for ^{210}Pb CIC modelling, reduce the associated uncertainties by increasing the number of data points and depth window, and select the right fingerprint tool that will help validate sedimentation rate results from modelling.

Sellafield-derived ^{137}Cs seemed to be a powerful fingerprinting tool for validating inshore and coastal BC sequestration rates estimated from ^{210}Pb CIC modelling, but more care must be taken when using this tool in offshore regions of the UK North Sea, mainly due to the presence of low concentrations of ^{137}Cs in this marine region and other potential ^{137}Cs sources. Other characteristic nuclear fingerprints might be able to help differentiate between the different sources and confirm the main source of ^{137}Cs present in the offshore UK North Sea.[21]

Anthropogenic ^{226}Ra seemed to be a good alternative fingerprinting tool for validating sedimentation rates in the direct vicinity of offshore oil & gas platforms in the North Sea. Nevertheless, further studies are needed to understand the fate of microparticles enriched in ^{226}Ra and its chemical analogues within the seabed better. In addition to its use as a fingerprinting tool, it could also help refine the knowledge of the ecological risk of produced water on benthic biota and, more generally, of the long-term environmental impact of this offshore industrial activity on the marine environment.[22,23]

5 ACKNOWLEDGEMENTS

The authors would like to thank Dr Ruth Parker (Cefas), Dr Natalie Hicks and Mr Hugo Woodward-Rowe (University of Essex, School of Life Sciences) for providing the sediment cores used in this paper as well as many colleagues involved in the preparation and analysis of sediment samples at Cefas.

6 REFERENCES

1. P. G. Appleby, W. Shotyk, A. Fankhauser. Lead-210 age dating of three peat bogs in the Jura Mountains, Switzerland. *Water Air Soil Pollut.*, 1997, 100, 223-231.
2. P. W. Swarzenski. In: *Encyclopedia of Scientific Dating Methods*, ed. W. Rink, J. Thompson, 2014, Springer, Dordrecht, 2014, 210Pb dating, 1-11. DOI: 10.1007/978-94-007-6326-5_236-1
3. A. Murdoch. Lake Ontario sediments in monitoring pollution. *Environ. Monit. Assess.,* 1993, 28, 117-129. DOI: 10.1007/BF00547030
4. I. Tolosa, J. M. Bayona, J. Albaigés. Aliphatic and polycyclic aromatic hydrocarbons and sulfur/oxygen derivatives in northwestern Mediterranean sediments: Spatial and temporal variability, fluxes, and budgets. *Environ. Sci. Technol.,* 1996, 30, 2495-2503. DOI: 10.1021/es950647x
5. A. B. MacKenzie, S. M. L. Hardie, J. G. Farmer, L. J. Eades, I. D. Pulford. Analytical and sampling constraints in ^{210}Pb dating. *Sci. Total Environ.,* 2011, 409, 1298-1304. DOI: 10.1016/j.scitotenv.2010.11.040
6. R. Garcia-Tenorio, M. Rozmaric, A. Harms, J. M. De Oliveira Godoy, M. Barsanti, A. Schirone, A. C. Ruiz-Fernández, J. A. Sanchez-Cabeza, P. McGinnity, M. Fujak, M. Eriksson, V. Hatje, A. Laissaoui, H. Q. Nguyen, E. Okuku, S. A. Al-Rousan, M. W. Yii, H. Heijnis, I. Osvath. From radiometry to chronology of a marine sediment core: A 210Pb dating interlaboratory comparison exercise organised by the IAEA. *Mar. Pollut. Bull.,* 2020, 159, 111490. DOI: 10.1016/j.marpolbul.2020.111490
7. A. Arias-Ortiz, P. Masqué, J. Garcia-Orellana, O. Serrano, I. Mazarrasa, N. Marbà, C. E. Lovelock, P. S. Lavery, C. M. Duarte. Reviews and syntheses: 210Pb-derived sediment and carbon accumulation rates in vegetated coastal ecosystems – setting the record straight. *Biogeosciences,* 2018, 15, 6791-6818. DOI: 10.5194/bg-15-6791-2018

8. O. Evrard, P-A. Chaboche, R. Ramon, A. Foucher, J. P. Laceby. A global review of sediment source fingerprinting research incorporating fallout radiocesium (^{137}Cs). *Geomorphology,* 2020*,* 362, 107103. DOI: 10.1016/j.geomorph.2020.107103
9. A. Foucher, P-A. Chaboche, P. Sabatier, O. Evrard. A worldwide meta-analysis (1977-2020) of sediment core dating using fallout radionuclides including ^{137}Cs and $^{210}Pb_{xs}$. *Earth Syst. Sci. Data.* 2021, 13, 4951-4966. DOI: 10.5194/essd-13-4951-2021
10. C. A. Graves, L. Benson, J. Aldridge, W. E. N. Austin, F. Dal Molin, V. G. Fonseca, N. Hicks, C. Hynes, S. Kroger, P.D. Lamb, C. Mason, C. Powell, C. Smeaton, S. K. Wexler, C. Would, R. Parker. Sedimentary carbon on the continental shelf: Emerging capabilities and research priorities for Blue Carbon. *Front. Mar. Sci.,* 2022, 9, 926215. DOI: 10.3389/fmars.2022.926215
11. OSPAR 2022 Modelling and assessment of additional concentrations of Naturally Occurring Radioactive Materials (NORM) in seawater from discharges of produced water from the offshore oil and gas sector in the North-East Atlantic. *OSPAR Commission Publication,* 925, 50pp.
12. H. De Haas, W. Boer, T. C. E. Van Weering. Recent sedimentation and organic carbon burial in a shelf sea: The North Sea. *Mar. Geol.* 1997, 144, 131-146. DOI: 10.1016/S0025-3227(97)00082-0
13. J. Robbins. In: *The biogeochemistry of lead in the environment,* ed. J. Nriagu, 1978, Elsevier, Amsterdam. Geochemical and geophysical applications of radioactive lead, 285-393
14. P. G. Appleby, F. Oldfield. The calculation of lead-210 dates assuming a constant rate of supply of unsupported ^{210}Pb to the sediment. *Catena*, 1978, 5, 1-8
15. S. Schmidt, H. Howa, A. Diallo, J. Martin, M. Cremer, P. Duros, C. Fontanier, B. Deflandre, E. Metzger, T Mulder. Recent sediment transport and deposition in the Cap-Ferret Canyon, South-East margin of Bay of Biscay. *Deep-Sea Res. II: Top. Stud. Oceanogr.* 2014, 104, 134-144. DOI: 10.1016/j.dsr2.2013.06.004
16. J. M. Dinsley, C. J. B. Gowing, A. L. Marriott. Natural and anthropogenic influences on atmospheric Pb-210 deposition and activity in sediments - a review. *British Geological Survey Internal Report* OR/17/047, 2018, 27pp
17. J. Gray, S. R. Jones, A. D. Smith. Discharges to the environment from the Sellafield site, 1951-1992. *J. Radiol. Prot.*, 1995, 15, 99
18. K. Perner, T. Leipe, O. Dellwig, A. Kuijpers, N. Mikkelsen, T. J. Andersen, J. Harff. Contamination of arctic Fjord sediments by Pb-Zn mining at Maarmorilik in central West Greenland. *Mar. Pollut. Bull.*, 2010, 60, 1065-1073. DOI: 10.1016/j.marpolbul.2010.01.019
19. Environment Agency, FSA, FSS, NIEA, NRW, SEPA Radioactivity in Food and the Environment. RIFE 26, 2020. Environment Agency, FSA, FSS, NIEA, NRW and SEPA, Bristol, London, Aberdeen, Belfast, Cardiff and Stirling, 2021.
20. F. Ahmad, K. Morris, G. T. W. Law, K. G. Taylor, S. Shaw. 2021 Fate of radium on the discharge of oil and gas produced water to the marine environment. *Chemosphere*, 2021, 273, 129550. DOI: 10.1016/j.chemosphere.2021.129550

21. M. Lin, J. Qiao, X. Hou, O. Dellwig, P. Steier, K. Hain, R. Golser, L. Zhu. 70-Year Anthropogenic Uranium Imprints of Nuclear Activities in Baltic Sea Sediments. *Environ. Sci. Technol.*, 2021, 55, 8918-8927. DOI: 10.1021/acs.est.1c02136
22. E. E. Cordes, D. O. B. Jones, T. A. Schlacher, D. J. Amon, A. F. Bernardino, S. Brooke, R. Carney, D. M. DeLeo, K. M. Dunlop, E. G. Escobar-Briones, A. R. Gates, L. Génio, J. Gobin, L-A. Henry, S. Herrera, S. Hoyt, M. Joye, S. Kark, N. C. Mestre, A. Metaxas, S. Pfeifer, K. Sink, A. K. Sweetman, U. Witte. Environmental Impacts of the Deep-Water Oil and Gas Industry: A Review to Guide Management Strategies. *Front. Environ. Sci.*, 2016, 4, 58. DOI: 10.3389/fenvs.2016.00058
23. D. J. Koppel, F. Kho, A. Hastings, D. Crouch, A. MacIntoch, T. Cresswell, S. Higgins. Current understanding and research needs for ecological risk assessments of naturally occurring radioactive materials (NORM) in subsea oil and gas pipelines. *J. Env. Rad.*, 2022, 241, 106774. DOI: 10.1016/j.jenvrad.2021.106774

CHAPTER 8 A MOLECULAR DYNAMICS STUDY OF HELIUM CLUSTERING IN HIGH TEMPERATURE PLUTONIUM DIOXIDE

Elanor Murray[a*], Ying Zhou[b], Peter Slater[a], Roger Smith[b], Pooja Goddard[b] and Helen Steele[c]

[a]School of Chemistry, University of Birmingham, Birmingham, UK, [b]School of Science, Loughborough University, Loughborough, UK, [c]Sellafield Ltd, Sellafield, Cumbria, UK

*elanormurray@gmail.com

1 INTRODUCTION

Nuclear power is set to play an essential role in the supply of reliable low-carbon energy for achieving urgent climate change mitigation.[1] The UK is committed to the use of nuclear energy to achieve its net-zero carbon emission 2050 target.[2] At present, the UK has the largest stockpile of civil separated Pu in the world, with circa 140 tonnes stored as PuO_2 powder at Sellafield, Cumbria.[3] This plutonium is a by-product of the nuclear fission of uranium from legacy reactors. The UK Government is developing a long-term solution for this material, with one option being re-use as mixed oxide (MOx) fuel in future reactors to supply new low-carbon energy. However, the implementation of this long-term solution will take decades, requiring the continued safe and secure storage of PuO_2 powder within inert nested steel storage cans.[3--5]

A key concern for the continued storage of the PuO_2 canisters is pressurisation, with radiogenic helium gas generation being a potential mechanism. Helium could diffuse to form large bubbles and at some point, be released from the lattice. Previous work proposed the main helium diffusion mechanism in PuO_2 to be via oxygen vacancy assisted inter-site hops with helium and oxygen having comparable diffusion rates. In contrast, plutonium vacancies have been found to be favourable trapping sites for helium.[6-8] However, plutonium vacancies and Frenkel pairs have been found to be energetically unstable and so are unlikely to be present in high concentrations.[9-15] Rather, it was predicted that helium clusters will occupy Schottky trio defect sites as these contain a stable plutonium vacancy.

A plateau in helium diffusivity was observed at high temperatures and this work aims to investigate contributing factors to such a plateau to determine if it is due to the clustering of helium as previously suggested.[6] This paper aims to provide further analysis of the high temperature region to investigate the helium accumulation into clusters and determine optimum helium bubble sizes.

2 METHODOLOGY

Interatomic potentials were used to describe the atomic interactions of helium in PuO_2. The many-body CRG potential was used for the Pu-Pu, Pu-O and O-O interactions.[16,17] To describe the He-He interactions, the Aziz potential was used.[18] A Lennard-Jones potential with parameters developed by Parfitt was used to describe the He-O interaction.[19] The He-Pu potential employed was the ZBL potential.[20] All calculations were performed in LAMMPS using an 8 x 8 x 8 supercell in the cubic fluorite crystal structure containing 6144 atoms with a box length of 43.1857 Å.[21]

Molecular dynamics (MD) techniques (in which the time-evolution of the system is followed) were used. The simulations were run in the isothermal-isobaric (NPT) ensemble with a Nose-Hoover thermostat and barostat.[22] Each system was relaxed via the conjugate gradient method then MD was run for 5 ns with a 1 fs timestep for a range of temperatures between 1000 and 3000 K.

OVITO was used to visualise and analyse the MD output files. Cluster analysis was performed to determine the formation of helium clusters.[23] A cluster is defined as a set of connected particles, each of which is within the (indirect) reach of the other particles in the same cluster. Thus, any two particles from the same cluster are connected by a continuous path consisting of steps that fulfil the selected neighbouring criterion. The neighbouring criterion was specified as a 2.2 Å interatomic distance.

3 RESULTS AND DISCUSSION

3.1 Helium clustering

A fluorite PuO_2 lattice with a helium concentration of 0.5 at.% was constructed by randomly inserting He atoms into Octahedral Interstitial Sites (OISs). No defects were initially present in the lattice. The system was evolved for 5 ns at 100 K intervals between 1000-3000 K. In all cases only a single He cluster formed and the number of helium atoms occupying the cluster at each temperature after 5 ns was determined and is plotted in Figure 1. It can be observed that no helium clusters formed below 2400 K, i.e., all helium atoms occupied isolated sites.

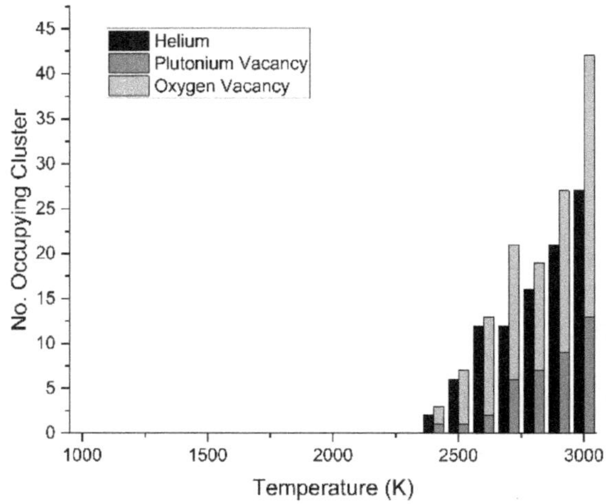

Figure 1 *Bar chart showing the number of helium atoms, plutonium vacancies and oxygen vacancies constituting the cluster at each temperature after 5ns.*

From Figure 1 it can be seen that the size of the cluster increases with increasing temperature. The number of plutonium and oxygen vacancies in the cluster is also displayed in Figure 1. It was also found that the number of vacancies in the cluster exceeds the number of helium atoms. This is in agreement with He:vacancy ratios found via dynamical calculations in previous work.[6]

Previous work studied the diffusivity of helium over the same temperature range (1000-3000 K).[6] From this it was found that below 1500 K helium exhibits no diffusive behaviour over 5 ns; helium atoms remain vibrating about their input OISs. Above 1500 K helium atoms were found to become mobile via oxygen vacancy assisted interstitial hops.[6] This is because above 1500 K, oxygen is mobile.[17] Upon visual inspection of the lattice evolution it was found that in the 1500 - 2300 K temperature range some helium atoms had moved to different OISs or were temporarily occupying oxygen vacancy sites. Further inspection revealed that helium was moving via oxygen vacancy assisted interstitial hops. However, though helium is mobile via oxygen vacancies in this temperature range, no helium clustering was present.

As displayed in Figure 1, above 2400 K helium clustering was observed. Previous diffusivity studies[6,17] found that above 2400 K, Pu can become momentarily mobile, forming temporary plutonium Frenkel pairs (PuFPs). Plutonium vacancies have been found to be favourable trap sites for helium in PuO_2.[6-8] However, isolated plutonium vacancies and PuFPs are energetically unfavourable defects in PuO_2.[9-15] So, as noted by

Neilson *et al.*,[7] such defects may not be available to trap helium since the occupation of an isolated Pu vacancy by a single helium atom does not prohibit PuFP recombination with the helium atom being ejected.[6] In our dynamical simulations, however, clustering does occur near sites where the Pu atom has been ejected. It is proposed that rather than a single Pu vacancy being the seed point for cluster growth, the formation of a Schottky trio (containing a stable Pu vacancy), or a region containing multiple defects could be the nucleation site for a helium cluster. Further dynamical simulations with a proper statistical analysis are required to justify fully this hypothesis.

Although helium atoms can diffuse at temperatures above 1500 K, it is suggested that it is the presence of plutonium vacancies that is essential for the nucleation of helium clusters. Thus, in initially defect free PuO_2, thermally generated Pu defects are present above 2400 K, so above this temperature helium clustering is viable.

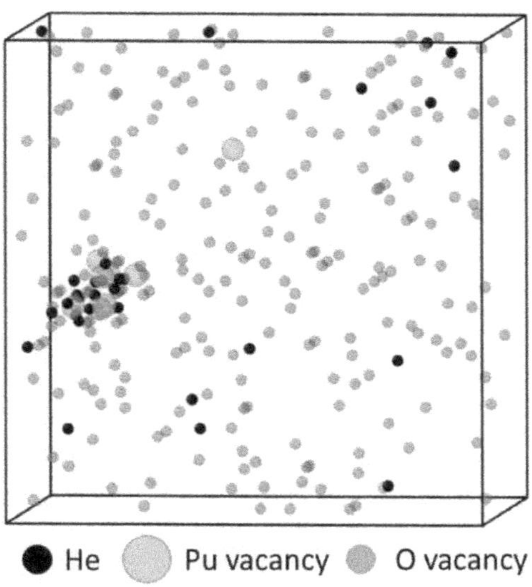

Figure 2 *Lattice with Pu and O ions hidden at 2700 K after 5 ns. A helium cluster has formed.*

Figure 2 displays the lattice at 2700 K after 5 ns, where it can be seen that a helium cluster has formed with plutonium and oxygen vacancies with a ratio of $He_{17}V_{24}$. In the lattice 39% of the helium atoms occupy the cluster, with the remaining helium atoms occupying OISs. Isolated Pu vacancies can be observed in Figure 2. However, it must be noted that these vacancies are transient phenomena, constantly forming and recombining.

In all the results so far only one helium cluster nucleated. In aged PuO_2 in which defects are present and distributed throughout the lattice, it could be expected that multiple smaller clusters may nucleate. The formation of multiple clusters was observed in systems with Schottky trios randomly distributed throughout the lattice.[6] In such work it was found that smaller clusters would nucleate and then merge to form clusters with lower He:vacancy ratios.

In the previous diffusivity study, above 2400 K a plateau in helium diffusivity was observed.[6] It is suggested that this may be due to the formation of helium clusters above this temperature, with the mobility of helium once trapped in the cluster significantly reduced.

3.2 He:vacancy ratio analysis

The number of helium atoms and vacancies in the cluster and the resultant He:vacancy ratio at each time step in a simulation is plotted for 2900 and 3000 K in Figure 3a and 3b, respectively. From this the evolution of the cluster can be analysed.

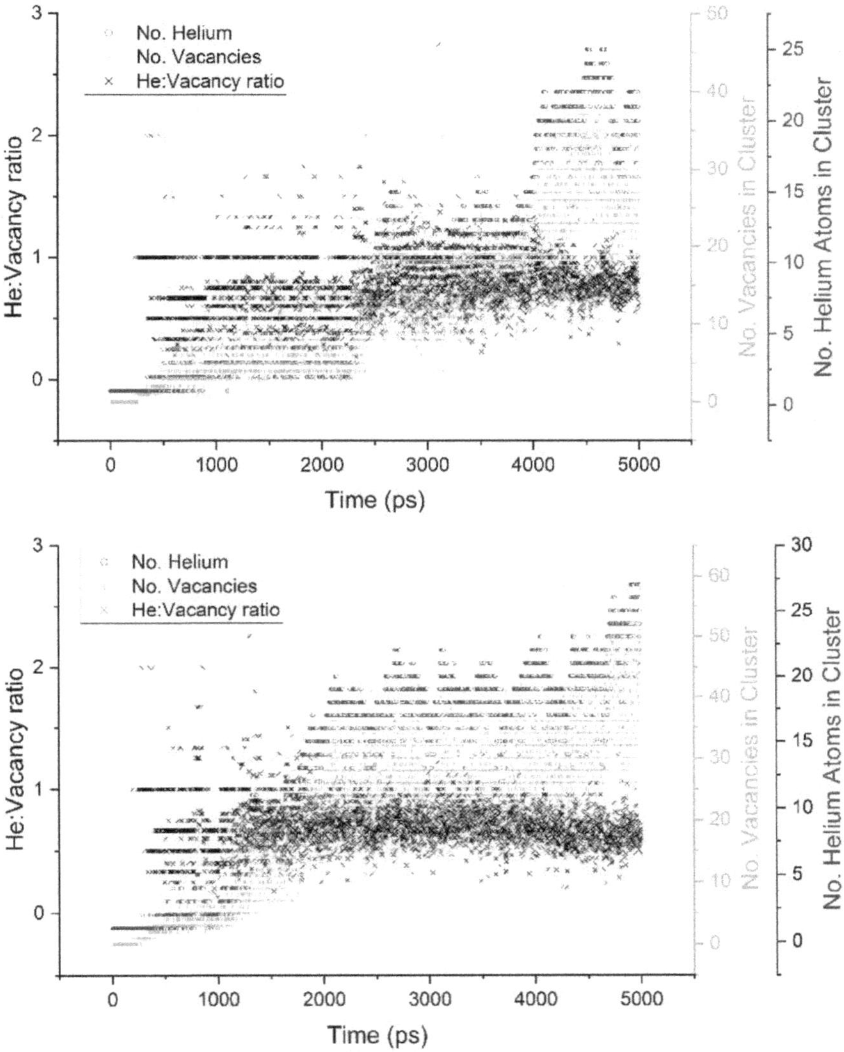

Figure 3 *(a) 2900 K. (b) 3000 K. Scatter plot with the three ordinate axes detailing the evolution of the He cluster as a function of time. The left axis (pink) plots the He:vacancy ratio for the cluster and the right axis (grey) plots the No. of vacancies in the cluster. The far-right axis (black) plots the No. of He atoms occupying the cluster.*

At both temperatures plotted in Figure 3 it can be seen that the number of helium atoms and vacancies in the cluster follow a similar trend. Both increase and decrease in tandem suggesting they are interdependent in the evolution of the cluster. After the cluster reaches a certain size, the average He:vacancy ratio remains roughly constant, at approximately 0.75:1. This is lower than the theoretical maximum ratio of ~ 3.5:1 previously determined statically,[6] implying that kinematic considerations could be a key factor in cluster evolution. In Figure 3b the number of helium atoms and vacancies displays a more gradual increase but the

He:vacancy ratio converges to approximately 0.75:1.

From Figures 3a and 3b it can be observed that the cluster grows more rapidly in the higher temperature lattice. A previously defined 'large' cluster (≥ 9 He[6]) forms after 2.5 ns in the lattice evolved at 2900 K and after 1.5 ns in the lattice evolved at 3000 K. The more rapid cluster evolution at the higher temperatures results from more thermally generated oxygen vacancies and therefore faster He diffusion.

The He:vacancy ratio was calculated at each time step for each lattice in which a large cluster had formed, and the mean He:vacancy ratio calculated. The 2500 K lattice was excluded due to the cluster only qualifying as 'large' in the last 400 ps. Figure 4 displays a box plot of the results. The range of the mean He:Vacancy ratios is narrow, between 0.88-0.71:1. It was found that the mean He:vacancy ratio decreases with increasing temperature. It is suggested that this trend may be due to the convergence in He:vacancy ratio values as the cluster grows in size, which is established earlier at higher temperatures (Figure 3). The overall mean He:vacancy averaged from each simulation was 0.8:1.

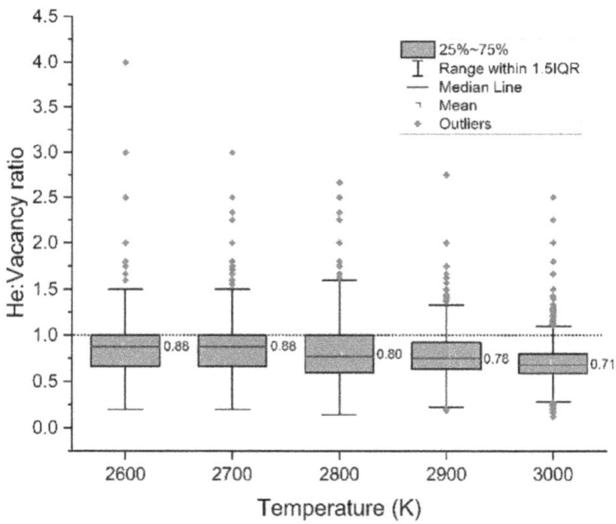

Figure 4 *Box plot of the He:vacancy ratio of the cluster at each temperature in which clustering was observed during 5 ns. The mean values are displayed.*

From Figure 4 it can be seen that at each temperature the interquartile range is below a He:vacancy ratio of 1:1, with the upper quartile being equal to 1 for the three lowest temperatures. Therefore, though clusters with ratios greater than 1:1 were found to be energetically feasible via energy minimisation studies,[6] they are predicted to be rare.

4 CONCLUSIONS

The clustering of helium in an initially defect free PuO_2 lattice has been observed at temperatures ≥ 2400 K. Although helium atoms are mobile via oxygen vacancies at temperatures above 1500 K, they do not cluster until plutonium vacancies are available. As such it is found that though helium atoms are mobile via interstitial hopping between 1500 - 2400 K, no clustering occurs at interstitial sites or oxygen vacancy sites over 5ns. Above 2400 K PuFP generation and annihilation is possible, leading to the availability of Pu vacancies which are helium trap sites.

Cluster sizes were found to increase with increasing temperature. In each simulation only one cluster formed. It has been found that a possible helium cluster nucleation site is a Schottky trio, as it contains a stabilised plutonium vacancy. It is suggested that in aged PuO_2, in which intrinsic defects are present, helium clustering may be possible at lower temperatures due to the availability of decay event generated Schottky trios. Additionally, in aged PuO_2, smaller helium clusters may nucleate throughout the lattice due to the distribution of intrinsic defects from collision cascades.

Cluster analysis performed on the generated clusters found that the average He:Vacancy ratio is 0.8:1. Thus, helium clusters forming in PuO_2 are predicted to contain more vacancies than helium atoms in contrast to the theoretical maximum value of 3.5:1 determined by static calculations.

5 CONFLICTS OF INTEREST

There are no conflicts of interest to declare.

6 ACKNOWLEDGEMENTS

The computations described in this paper were performed using the University of Birmingham's BlueBEAR HPC service, which provides a High-Performance Computing service to the University's research community. This project was funded by the Nuclear Decommissioning Authority and carried out as part of the EPSRC's TRANSCEND project (EP/S01019X/1).

7 REFERENCES

1. IAEA, *The Potential Role of Nuclear Energy in National Climate Change Mitigation Strategies*, 1984, 2021.

2. BEIS, *Net Zero Strategy: Build Back Greener*, 2021.

3. NDA, *Progress on Plutonium Consolidation, Storage and Disposition*, 2019.

4 N. Hyatt, *Energy Policy*, 2017, **101**, 303.

5. N. Hyatt, *npj Materials Degradation*, 2020, **4**, 28.

6. E. Murray, Y. Zhou, P. Slater, R. Smith, P. Goddard and H. Steele, *Physical Chemistry Chemical Physics,* 2022, **24**, 20709.

7. W Neilson, H. Steele, N. Kaltsoyannis and S. Murphy, *Physical Chemistry Chemical Physics*, 2022, **24**, 8245.

8. E. Murray, N. Palmer, W. Neilson, R. Jackson, M. Read, H. Steele and R. Orr, Presented in part at the WM Symposia, Phoenix, AZ, March, 2022.

9. W. Neilson, J. Pegg, H. Steele and S. Murphy, *Physical Chemistry Chemical Physics,* 2021, **23**, 4544.

10. X. Tian, T. Gao, C. Lu, J. Shang and H. Xiao, *The European Physical Journal B*, 2013, **86**, 179.

11. P. Tiwary, A. van de Walle, B. Jeon and N. Grønbech-Jensen, *Physical Review B*, 2011, **83**, 094104.

12. M. Freyss, N. Vergnet and T. Petit, *Journal of Nuclear Materials*, 2006, **352**, 144.

13. S. Hernandez and E. Holby*, The Journal of Physical Chemistry C,* 2016, **120**, 13095.

14. M. Read, S. Walker and R. Jackson, *Journal of Nuclear Materials,* 2014, **448**, 20.

15. L-F. Wang, B. Sun, H-F. Liu, D-Y. Lin and H-F. Song, *Journal of Nuclear Materials*, 2019, **526**, 151762.

16. M. Cooper, M. Rushton and R. Grimes, *Journal of Physics: Condensed Matter,* 2014, **26,** 105401.

17. M. Cooper, S. Murphy, M. Rushton and R. Grimes, *Journal of Nuclear Materials*, 2015, **461**, 206.

18. R. Aziz, A. Janzen and M. Moldover, *Physical Review Letters*, 1995, **74,** 1586.

19. D. Parfitt and R. Grimes, *Journal of Nuclear Materials*, 2008, **381**, 216.

20. J. Ziegler and J. Biersack, *The Stopping and Range of Ions in Matter*, Springer US, Boston, MA, 1985.

21. A. Thompson, H. Aktulga, R. Berger, D. Bolintineanu, W. Brown, P. Crozier, P. in 't Veld, A. Kohlmeyer, S. Moore, T. Nguyen, R. Shan, M. Stevens, J. Tranchida, C. Trott and S. Plimpton, *Computer Physics Communications*, 2022, **271,** 108171.

22. B. Todd and P. Daivis, Nonequilibrium Molecular Dynamics: Theory, Algorithms and Applications, Cambridge University Press, Cambridge, 2017.

23. A. Stukowski, *Modelling and Simulation in Materials Science and Engineering*, 2010, **18.**

CHAPTER 9 FAST Am, Pu AND Sr SEPARATION AFTER AUTOMATED FUSION OF HIGHLY DENSE BARITE CONCRETE

M. Jäggi[a*], M. Heule[a], F. Köhler[a], N. Walter[a] and S. Mayer[a]

[a]Paul Scherrer Institute, Department of Radiation Safety and Security, Radioanalytics, CH-5232 Villigen, Switzerland

*maya.jaeggi@psi.ch

1 INTRODUCTION

The borate fusion process is capable dissolving hard to dissolve complex oxides, i.e. slags and ores[1], and geological material[2]. Croudace et al.[3] were first to apply borate fusion in the context of radionuclide analysis, to be best to our knowledge. Thereafter, borate fusion became a standard method to dissolve various matrices completely for radionuclide analysis.[4-9] Sample amounts of up to 5 g can be fused in crucibles of 25–30 ml volume (Pt/Au 95 %/5 %), when the process is done manually.[3,5,10] For secure handling of the fusion process, automated borate "fluxer" systems ensure process consistency and ease of use to dissolve barite concrete or similar samples.[9] Commercially available automatic borate "fluxer" – systems in use are for example from Malvern Panalytical (Kassel, Ge), such as the Claisse LeNeo[9] or from SPEX (Metuchen, USA), such as the Katanax K2 Prime.[11] Using the commercial borate "fluxers",[12] an array of sample matrices could be dissolved, such as concrete, cement, soil, graphite, and sediment samples,[11] or in addition, also mussels and seaweed.[7]

For the automatic "fluxers", it is usually mentioned that only small amounts of material, i.e. less than 1 g, can be fused when using the Pt/Au crucibles with 25–30 ml volume. This is most likely due to the requirement that the hot melt in the crucible has to be thoroughly mixed by high rocking angles and speed. However, for some samples containing low levels of radioactivity it is highly important to increase the sample mass to obtain correspondingly low detection limits.

Previous work addressed the volume limitations. Trdin et al.[7] fused up to 2 g of 6 different sample matrices with a Claisse LeNeo, and Braysher et al.[11] adjusted sample to flux ratios and adapted the parameters for fusion with the Katanax K2 Prime fluxer (SPEX, Metuchen, USA), which allowed for dissolution of up to 5 g of material.

Sequential separation of radionuclides of a single sample has been shown earlier.[13] Due to the development of transuranium-specific extraction chromatography (EC) materials,[14] the SR Resin,[15] and the DGA resin to pre-concentrate tri- and tetra valent actinides, i.e. americium,[16] multi-radionuclide separation of a single sample became faster.[17] For the fast and simultaneous determination of hard-to-detect radionuclides in water and swipe samples, the TEVA-UTEVA-DGA-SR Resin-TRU were stacked in series to separate U/Pu from TEVA, Th from UTEVA, Am/Cm from DGA, Sr from SR Resin and Fe from TRU.[17]

The aim of the current study was to reduce the 5 day process time for fusion and chemical separation of 0.3 g – 3 g of highly dense barite concrete, and to improve the Am, Pu and Sr tracer recoveries to exceed a 70 % threshold, compared to earlier work (method 3).[8] To achieve the proposed goals, we compared two new methods for our laboratory, method 1, loose resins (Triskem international, Bruz, France) UTEVA-TEVA-DGA-SR Resin with a fixed UTEVA column, method 2, cartridges (Triskem international, Bruz, France) UTEVA-TEVA-DGA-SR Resin to method 3,[8] DGA, Bio-Rad AG 1-X2 (Bio-Rad Laboratories, Kabelsketal, Ge)-UTEVA, DGA, TEVA, SR Resin (Triskem international, Bruz, France).

2 MATERIALS AND METHODS

2.1 Fusion and separation of Am, Pu, Sr (U, Th,) in 0.3 g of barite concrete

Earlier work, method 3, presented a 5 day manual fusion process and chemical separation scheme for barite concrete.[8] To reduce work labour and to improve Am, Pu and Sr tracer recoveries above the threshold of 70 %, two new separation schemes, method 1 and 2 are were developed in our laboratory (Figure 1). The automated fusion process and the individual separation steps are explained in the next sub-sections.

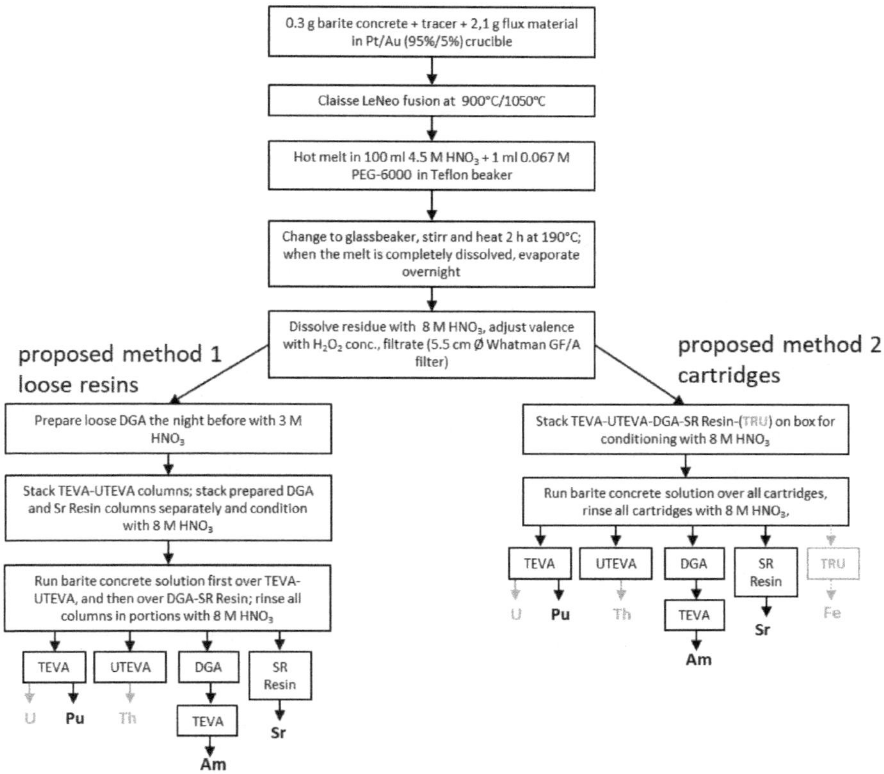

Figure 1 *Fusion process and separation scheme for Am, Pu, Sr (optionally U, Th) in barite concrete. Method 1 works with loose resins, and method 2 works with cartridges based on Dai et al.*[17]

2.2 Evaluation of three different flux materials

Three different flux agents were tested for best fusion performance. To reduce the amount of residual melt left in the crucibles, the addition of the non-wetting agent lithium-bromide was tested. Lithium-bromide reduces the surface tension of the hot melt, which reduces the problem of residues sticking to the crucible walls.[18] When increasing the LiBr content, less remainders of the melt were sticking to the crucibles using F3, as long as the crucibles kept a smooth surface finish (Table 1).

Table 1 *Three different flux materials.*

Abbreviation	Flux agent	Ratio
F1	$LiBO_2$ / $Li_2B_4O_7$	80 %/ 20 %
F2	$LiBO_2$ / $Li_2B_4O_7$ / LiBr	49.5 % / 48.5 % / 1 %
F3	$LiBO_2$ / LiBr	98.5 % / 1.5 %

2.3 Preparation for fusion of 0.3 g – 3 g of inactive barite concrete and 0.3 g – 0.5 g of sludge samples

The pre-set parameter file ("solution - high refractory") of the Claisse LeNeo was used for 0.3–0.5 g of dry-ashed (600 °C) inactive barite concrete and radioactive sludge samples. The temperature was set of 900 °C and was maintained for 1 h before fusion. The total mixture of material and flux agent should not exceed 2.5 g, therefore 2.1–1.9 g of flux agent F1 was added to the corresponding sample weights. The tracers ^{243}Am (50 mBq), ^{242}Pu (50 mBq) and ^{85}Sr (10 Bq for LSC-Cherenkov) were added to the Pt/Au crucible (95 %/5 %). In case U and Th were measured, 50 mBq ^{232}U tracer was added to one of the two needed aliquots, according to Dai et al.[17] The crucible was heated at 150 °C for 1-2 h to allow for the acid of the tracers to evaporate. As shown in the Section 2.2, F1 $LiBO_2/Li_2B_4O_7$ (F1: 80 %/20 %), was henceforth used as flux agent. 100 ml of 4.5 M HNO_3 was prepared in a 150 ml Teflon beaker with magnetic stirrer, and 1 ml 0.067 M PEG-6000 (PEG = polyethylene glycol, Sigma-Aldrich) was added for SiO_2-gel flocculation.

When fusing 3 g of dry-ashed (600 °C) inactive barite concrete, adaptations had to be made to the parameter file ("solution – high refractory") and to the amount of flux material added to 3 g of inactive barite concrete, where adaptations can be seen in Table 2.

Table 2 *Comparison of pre-set and new parameters for lithium borate fusion.*

Procedure	**Pre-set**	**New**
Mass of barite concrete [g]	0.3	3.0
Mass of flux agent F1 [g]	2.1	3.6
Flux agent F1/barite concrete ratio	1:7	1:1.2
Preheating/Fusion temperature [°C]	900 / 1050	1050 / 1065
Rocking angle [deg]	20 - 45	5
Rocking speed [rpm]	5 - 30	5
Acid solution: HNO_3	100 ml 4.5 M	100 ml 4.5 M
SiO_2 flocculation: 0.067 M PEG-6000	1 ml	2 ml
Temperature/time of heating after fusion	190 °C/1 h – 2 h	190 °C/3 h

In order to deal with the total mass increase from 2.4 g to 6.6 g of barite concrete and flux agent F1 mixture, the total fusion time was increased from 22.5 min to 30.5 min to ensure a complete fusion process.

2.4 Automated fusion with the Claisse LeNeo

The fusion process of 0.3–0.5 g matrix/3 g matrix was started with a maximum temperature of 1050 °C/1065 °C. The procedure took 22.5/30.5 min. After 19/27 min the hot liquid melt was poured directly into 100 ml of 4.5 M HNO_3 at room temperature, followed by 3.5 min of active cooling by two built-in air fans. After the fusion, the solution of the Teflon beaker was transferred into a 250 ml glass beaker. In case there was some melt material left inside the crucible, the remaining melt was heated with 10 ml of 4.5 M HNO_3 at 190 °C separately. In parallel, the 250 ml glass beaker was heated at 190 °C while stirring for 1–2 h/3 h until the glass was completely dissolved. Then both solutions were combined in the 250 ml beaker and evaporated overnight. The dried residues were then dissolved in 15 ml of 8 M HNO_3. The sample was placed on a hot plate and 1 ml of concentrated H_2O_2 was added for adjustment of Pu oxidation state to +IV. The sample was heated to 150 °C, left to react for 0.5 h at that temperature. The 8 M HNO_3 solution was left to cool down to room temperature, enabling the SiO_2-gel flocculation to complete. The sample was filtrated with a Whatman GF/A 5.5 cm diameter filter (Sigma Aldrich), to remove the flocculated SiO_2-gel. Rinsing was limited to a maximum of 35 ml of 8 M HNO_3, since the amount of solution had an impact ^{85}Sr recovery, when separating with the SR Resin. Apparently, addition of volumes greater than 50 ml resulted in a slow decomposition of the SR Resin, releasing previously bound ^{85}Sr / ^{90}Sr.The filtrated solution was further processed, see sections 2.5 and 2.9.

2.5 Proposed chemical separation method 1 with loose resins

The loose resins TEVA, DGA and SR Resin were prepared in columns, together with a pre-packaged UTEVA column (resins from Triskem international, Bruz, France). DGA was prepared one night in advance by placing 0.6 g of the loose resin in a Bio-Rad econo-column (1 cm Ø / 20 cm length). A Whatman GF/B (Sigma Aldrich) 1 cm diameter filter pad was placed on the resin bed. Twice 5 ml of 3 M HNO_3 was run through the column for conditioning. The stopcock was closed and another 5 ml of 3 M HNO_3 was added to wet the loose DGA resin overnight. For the TEVA, 0.5 g loose resin was weighed into a Bio-Rad-column (Poly-Prep, Bio-Rad Cat. Nr. 731-1550) and was conditioned with some demineralized H_2O. Using a fine glass rod the air bubbles from the resin were eliminated. The covers of the UTEVA column were removed and the storage solution was emptied completely by pushing the frit on the resin bed. The TEVA column was directly placed above the UTEVA column and both were conditioned with 15 ml of 8 M HNO_3 (waste). 1 g of SR Resin was weighed in a Bio-Rad econo-column (1 cm diameter / 20 cm length). A Whatman GF/B (Sigma Aldrich) 1 cm diameter filter pad was placed on the resin bed. The resin was rinsed with 2-3 ml of 0.05 M HNO_3. The 3 M HNO_3 wetting solution of the DGA was run through the resin, and the DGA column was placed over the SR Resin column and both were conditioned with 15 ml of 8 M HNO_3 (waste).

The final barite concrete solution (Section 2.4) was loaded in portions over the TEVA-UTEVA columns and collected in a 50 ml glass beaker. This solution was then loaded in portions over the DGA-SR Resin columns and collected for waste in another 50 ml glass beaker. The first 50 ml glass beaker was rinsed with 10 ml of 8 M HNO_3 which were fed through all 4 columns. Additional 5 ml of 8 M HNO_3 rinsing solution was loaded onto all 4 columns directly. Then the columns were all separated for radionuclide elution. As long as the DGA column had still a yellow tint from the Fe^{3+}, then it was rinsed separately with 5 ml of 8 M HNO_3 in several steps until the tint vanished and the column reached a uniform colour impression.

Sr was eluted first from the SR Resin with 10 ml of 0.05 M HNO_3 into a 20 ml scintillation vial for liquid scintillation (LSC) -Cherenkov counting. Thus, the ^{90}Y ingrowth was measured, and ^{85}Sr recovery was determined by gamma-spectrometry. U had to be eluted from the TEVA first with 10 ml concentrated HCl, before Pu elution, to prevent the elution of both radionuclides together. However, the U solution was discarded. Pu was eluted from the TEVA with 20 ml of 9 M HCl / 0.01 M HI into a 25 ml glass beaker containing 0.6 ml of 1 M $NaHSO_4$ and 0.6 ml of concentrated H_2SO_4. The mixture was evaporated and calcined with at least 500 °C. The residue was dissolved in a buffer solution, and electrodeposited according to Bajo and Eikenberg.[19] The activity concentration of Pu as determined by alpha-spectrometry. The UTEVA column was discarded, since the Th activity concentration was not determined. Finally, Am was eluted from the DGA resin with 20 ml of 0.05 M HCl in a 50 ml glass beaker and the Am eluate was evaporated. The residue was twice dissolved in 1 ml of concentrated HNO_3 and evaporated to remove the organic matter. Then the residue was once dissolved with 1 ml of concentrated HCl and evaporated until dry. Am was separated from the lanthanides, by using a TEVA resin as explained in Section 2.11.

2.6 Proposed chemical separation method 2 with cartridges

The TEVA, UTEVA, DGA, SR Resin cartridges were stacked together in this order (resins from Triskem international, Bruz, France). A 25 ml syringe as reservoir was mounted on top of the TEVA cartridge and a flow regulation valve below the last cartridge was installed. The stacked resins were placed on top of a vacuum box. Inside the box, the outlet was extended using a Teflon tube which led to a 50 ml PE collection reservoir. The vacuum box was sealed tightly and the flow regulation valves were closed. 15 ml of 8 M HNO_3 were placed in the 25 ml syringe reservoir, and the flow regulation valves were opened and the vacuum of 5 mbar – 10 mbar was applied, to ensure a maximum droplet flow of 1 ml per minute. Thus, the cartridges were conditioned, the solution discarded, and the flow regulation valves were closed again.

The filtrated 8 M HNO_3 solution (see Section 2.4)) was added in portions to the 25 ml top reservoir. This solution was drawn through the stack of cartridges under vacuum, where the waste solution was discarded. Thus Am, Pu and Sr were captured on the corresponding resins. 10 ml of 8 M HNO_3 were used to rinse the beaker and the cartridges. Additional 5 ml of 8 M HNO_3 were used to rinse the cartridges only. The 8 M HNO_3 rinsing solution was discarded as waste, and the cartridges were separated. Sr was eluted directly from the SR Resin with 10 ml of 0.05 M HNO_3 into a liquid scintillation vial for LSC-Cherenkov measurement. The starting time and date of the elution were noted, since the ^{90}Y ingrowth started as soon as the rinsing with 8 M HNO_3 had stopped.

The DGA cartridge was placed on the vacuum box, and rinsed several times in steps using 5 ml of 8 M HNO_3, until the resin appeared in even colouring. Finally, Am was eluted with 20 ml of 0.05 M HCl, poured into a 50 ml glass beaker and evaporated until dry. The Am residue was dissolved and evaporated twice in 1 ml concentrated HNO_3 to destroy organic matter from the DGA cartridge. Then it was once dissolved and evaporated in 1 ml concentrated HCl. Am was further purified from the lanthanides using a TEVA resin, as explained in section 2.11.

The U was eluted with 10 ml concentrated HCl as waste solution. Pu was subsequently eluted with 20 ml of 0.1 M HCl and poured directly into a 25 ml glass beaker[8] containing 0.6 ml of 1 M $NaHSO_4$ and 0.6 ml of concentrated H_2SO_4. The solution was evaporated and calcined at 500 °C. The residue was dissolved in a buffer solution and electrodeposited.[19] Pu activity concentrations were measured by alpha-spectrometry. The UTEVA cartridge was discarded, since the Th activity concentration was not measured.

2.7 Am, Pu, and Sr fusion and chemical separation, method 3

Fusion was done manually for 1 g of dry-ashed inactive barite concrete with 7 g of F1 flux agent and Am, Pu, and Sr tracers. The hot melt was poured into a 250 glass beaker containing 150 ml of 4.5 M HNO_3 and 1 ml of 0.2 M PEG-2000. After the melt was dissolved completely, the sample solution was stirred overnight at 60 °C. Then the solution was filtrated with a Whatman GF/A filter of 5.5 cm diameter (Sigma Aldrich) and evaporated until dry. The residue was dissolved in 6 M HCl and the actinides preconcentrated using a loose DGA resin. The two-valent Sr flowed through and the solution was evaporated. Sr was further processed by Ca-oxalate precipitation and separated on a SR Resin. Pu was separated using a Bio-Rad AG 1-X2. U and Th were discarded (UTEVA), and Am was purified using a DGA cartridge. Finally, the lanthanide separation was done according to Section 2.11. This fusion and separation method is explained in more detail in Jäggi et al.[8]

2.8 Am separation from the lanthanides for 0.3–3 g of inactive barite concrete

An amount of 0.5 g TEVA was prepared in a column (Poly-Prep BIO-RAD Cat. 731-1550) and rinsed with 2-3 ml of demineralized H_2O. Conditioning was done with 5 ml of 2 M NH_4SCN–0.1 M HCOOH. The evaporated Am fraction (see Sections 2.5, 2.6, and 2.7) was re-dissolved in 5 ml of 2 M NH_4SCN–0.1 M HCOOH. This solution was loaded on the TEVA resin. The glass beaker and the resin were rinsed with 5 ml of 2 M NH_4SCN. The glass beaker and the resin were then rinsed with 10 ml of 1 M NH_4SCN. Subsequently, Am was eluted with 20 ml of 2 M HCl into a 25 ml glass beaker containing 0.6 ml of 1 M $NaHSO_4$ and 1 ml of concentrated H_2SO_4. The sample was evaporated and calcined at

500 °C. The residue was dissolved in a buffer system, electrodeposited[19] and measured by alpha-spectrometry.

2.9 Radioactive sludge samples for validation of the automated fusion process and chemical separation

For validation of the automated fusion process and chemical separation, 19 radioactive sludge samples were used. These sludge samples were part of a decommissioning project. The samples originated from 4 different detention ponds (Paul Scherrer Institute). After centrifugation, the samples were dry-ashed in steps of 40 °C, 110 °C, and 600 °C. For validation, samples dry-ashed at 600 °C were chosen.

3 RESULTS AND DISCUSSION

Based on the chosen matrix, the flux agent for fusion varies.[11] A mixture of $LiBO_2$ (melting point 849 °C) and $Li_2B_4O_7$ (melting point 917 °C) flux agents,[20] together with method 1 achieved best ^{243}Am and ^{85}Sr tracer recoveries (Table 3). Pure lithium metaborate ($LiBO_2$: F3) is excellent to fuse silicic acid to 100 %,[20] maybe explaining the high ^{242}Pu recovery using F3. However, when regarding the lower ^{242}Pu recovery shown by applying flux agent F3 and separation technique method 2 (Table 3), not only the flux agent, but also the separation technique may have influenced the tracer recoveries. Therefore the flux agent F1, as proposed by Bennet and Oliver,[20] despite the remaining melt pieces in the crucible after fusion, was chosen for further fusion of barite concrete.

Table 3 *^{243}Am, ^{242}Pu and ^{85}Sr recoveries depending on method and/or flux material for inactive barite concrete.*

Flux agent	**Tracer recoveries [%]**		
	^{243}Am	^{242}Pu	^{85}Sr
F1, method 1	93 ± 4	76 ± 7	78 ± 6
F2, method 1	81 ± 9	89 ± 6	70 ± 6
F3, method 1	85 ± 10	91 ± 6	74 ± 2
F3, method 2	73 ± 18	79 ± 3	66 ± 6

In general, the handling of the method 1 was more tricky and laborious than method 2. This was due to the two 20 cm columns containing the loose resins DGA and SR Resin. Furthermore, since only twice 2 columns were in series, more rinsing steps with 8 M HNO_3 were necessary in method 1, compared to method 2, making it more time consuming. In addition, chemical separation in method 1 was done by gravity and method 2 by vacuum. Additional differences between the method 1 and 2 are the amounts of resins as shown in Table 4. Except for the pre-packaged UTEVA resin whose quantity was fixed, lower quantities of loose resin was used for the TEVA and DGA. Concerning the SR Resin, higher quantities of loose resin were applied compared to the cartridge. However, the quantity of loose resin can be adapted, which is an advantage of method 1.

Table 4 *Differences in resin amount between loose resins and pre-packaged columns and cartridges.*

Triskem resin	Method 1 loose resin [g]	Method 2 cartridges [g]
TEVA	0.5	0.7
UTEVA pre-packaged/cartridge	0.78	0.78
DGA	0.6	0.74
SR Resin	1	0.66

Fusion with flux agent F1 was chosen for the separation techniques method 1 – 3. The proposed method 1 required 3.5 days, since the chemical separation with loose resins was quite time consuming. Method 2 was the fastest, where the automated fusion process and chemical separation could be completed within 2.5 days – 3 days. The duration of the fusion process and the radionuclide separation for the earlier published method 3 lasted 5 days.[8] This was due to the laborious Ca-oxalate precipitation step before Sr separation and due to the longer procedure for the manually operated fusion of the inactive barite concrete.

Table 5 *Comparison of the tracer recoveries for inactive barite concrete[8] of the two methods 1 and 2 (0.3 g), and the earlier published method 3 (1 g), using F1 as flux agent, n = 6.*

Radioactive tracer	**Method 1, recovery [%]**	**Method 2, recovery [%]**	**Method 3, recovery [%]**
^{243}Am	85 ± 13	87 ± 6	63 ± 16
^{242}Pu	84 ± 14	92 ± 3	86 ± 2
^{85}Sr	74 ± 8	74 ± 16	66 ± 3

Comparing the tracer recoveries for methods 1 - 3 of the same inactive barite concrete sample, as for method 3,[8] it can be seen that method 2 shows the highest recoveries for ^{243}Am and ^{242}Pu. In addition, method 2 performs similarly to method 1 for the ^{85}Sr recoveries (table 5), also exceeding the set 70 % threshold value. Using the flux agent F1 for the automated fusion process and the cartridges for radionuclide separation of barite concrete, we achieved the best tracer recoveries. In addition, by applying method 2, a two day reduction for the fusion and chemical separation was achieved compared to earlier work.[8]

To reduce the detection limits for Am, Pu, and Sr activity concentrations, more barite concrete had to be fused. Applying the automated fusion process with flux agent F1, to dissolve 3 g of inactive barite concrete, fusion parameters had to be adapted. Therefore, the flux-to-sample ratio was adapted, the set- and fusion temperatures were increased, and the rocking angle and speed of movement were lowered similar to the procedures used by Braysher et al.[11] To deal with the total mass increase from 2.4 g to 6.6 g of barite concrete and flux agent F1 mixture, the total fusion time was increased from 22.5 min to 30.5 min to ensure a complete fusion process.

The chemical separation scheme method 2 also had to be adapted for the ^{85}Sr tracer recoveries (Table 6). The recoveries of ^{243}Am and ^{242}Pu were not affected by the matrix increase to 3 g. After the outlier test of Grubbs, recoveries of 95 % ± 5 % (original n = 12, after Grubbs test n = 11) for ^{243}Am and 80 % ± 5 % (no outlier, n = 14) for ^{242}Pu were achieved, respectively.

Table 6 *Improvement steps for the fusion and radionuclide separation of 3 g inactive barite concrete, spiked with ^{85}Sr (^{243}Am and ^{242}Pu), n = 2. When filters were not measured (n.m.), the total ^{85}Sr recovery could not be summed (-).*

Tracer recovery / Improvement steps n = 2	**^{85}Sr [%]**	**^{85}Sr [%] filter 1**	**^{85}Sr [%] filter 2**	**^{85}Sr [%] total**
step 1: method 2 F1	27 ± 3	n.m.	n.m.	-
step 2: manual fusion	27 ± 3	n.m.	n.m.	-
step 3: 2 g loose SR resin	49 ± 6	n.m.	37 ± 2	86 ± 6
step 4: 2.5 g loose SR resin, two step filtration day 1+2	70 ± 7	9.3 ± 0.2	6.6 ± 2.0	86 ± 7
step 5: step 4 + 2 ml PEG-6000	71 ± 6	10.1 ± 0.1	3.7 ± 3.3	85 ± 7
step 6: step 5 + 100 mg SiO_2 (quartz)	66 ± 4	8.9 ± 0.3	8.7 ± 1.8	83 ± 4
step 7: see step 5, stirring stop after hot melt dissolution	74 ± 2	n.m.	n.m.	-

With step 1 we tested, if 3 g of inactive barite concrete could be separated by method 2 after fusion with F1 (Table 6). Less than 30 % recovery was found for ^{85}Sr. The same procedure was applied in step 2, yet the fusion was done manually, where fusion could be observed[8]. No spill-over was detected, but ^{85}Sr recovery was still below 30 %.

Two effects were identified which could be responsible for the loss of more than 70 % of the ^{85}Sr tracer. Such as, a) Sr precipitation in the form of $SrSO_4$ together with SiO_2-gel flocculation could occur, and/or b), the Sr capacity (8 mg/ml SR Resin)[15] of the SR Resin was probably exceeded. The SR Resin amount was increased to 2 g loose resin, instead 0.66 g (cartridge) in step 3. Tracer recovery increased to 49 %. However 37 % were lost on the filter 2. With step 4, 2.5 g of loose SR Resin were applied together with a two-step filtration. Filtration 1 was done on day 1 after fusion. The solution was evaporated overnight. Filtration 2 was done on day 2 after the residue was dissolved in 8 M

HNO_3. Step 4 led to 70 % of ^{85}Sr recovery and to a limited loss of 16 % on filter 1 and 2.

With steps 5 and 6, the enhancement of SiO_2-gel flocculation was tested, since flocculation may take longer than 12 h – 24 h[21]. The experiment was repeated up to step 4, but flocculation was carried out using 2 ml of 0.067 M PEG-6000 in step 5. The resulting ^{85}Sr recovery was 71 %, and the loss on the filter was reduced to 14 %. Another option in step 6 was the addition of 100 mg SiO_2 in the form of quartz-sand to further enhance the SiO_2-gel flocculation.[21] The ^{85}Sr loss increased to 18 % and the recovery was slightly reduced to 66 %. Step 5 was repeated in step 7 with one variation, after dissolution of the hot melt in the acid solution, stirring was stopped to enhance SiO_2-gel flocculation.[21] This variation resulted in the highest ^{85}Sr recovery rate of 74 %.

Finally, the proposed automated fusion process with flux agent F1 and adjoining chemical separation scheme method 2 were validated with the tracer recoveries of 19 radioactive sludge samples (Table 7). Besides gamma measurements, alpha- and beta-emitters were of high interest, thus the recoveries of the tracers ^{243}Am, ^{242}Pu, and ^{85}Sr were investigated. Based on the activity concentrations measured by gamma-spectrometry, 0.3 g to 0.5 g of dry-ashed sludge samples were fused together with 2.1 g to 1.9 g of F1 flux agent, respectively. The fusion and chemical radionuclide separation for up to 4 samples could be achieved within 2.5 days to 3 days. The average recoveries for all determined radioactive tracers are excellent as given in Table 7.

Table 7 *Tracer recoveries for several radioactive sludge samples from retention basins at PSI are shown. For the averaged results, the standard deviation is shown.*

Tracer recovery	**^{243}Am [%] n = 19**	**^{242}Pu [%] n = 19**	**^{85}Sr [%] n= 19**
Average	69 ± 17	84 ± 8	72 ± 7
Median	70	80	74
Maximum	98	99	82
Minimum	44	62	51

4 CONCLUSION

The secure automated fusion process was successfully implemented in our laboratory to fuse 0.3 g – 3 g of barite concrete This allows for enhanced throughput of hard to be dissolved matrices. The best procedure, method 2 and automatic fusion with flux agent F1, led to the highest averaged ($n = 6$) tracer recoveries for ^{243}Am (87 % ± 6 %), ^{242}Pu (92 % ± 3 %), and ^{85}Sr (74 % ± 16%) for inactive barite concrete, exceeding the set threshold of 70 % tracer recovery. Additionally, a reduction of analysis time by 2 days was successfully demonstrated compared to earlier work.[8] The best procedure was validated by the excellent results for Am, Pu and Sr of 19 radioactive sludge samples. To reduce the detection limits for alpha/beta activity concentrations, 3 g of barite concrete was successfully fused by adapting the flux-to-sample ratio, increasing the set- and fusion temperatures, and by reducing the rocking angle and speed of movement. A maximum ^{85}Sr recovery of 74 % was achieved by increasing the PEG-6000 quantity, by applying a two-step filtration, by adapting the stirring time before filtration[21] and by adjusting the amount of loose SR Resin to the total Sr content of the inactive barite concrete.

5 ACKNOWLEDGEMENTS

The authors thank Silvia Köchli for the ICP-OES measurements, and they thank for the support from the Swiss Nuclear Safety Inspectorate (ENSI), contract number: CTR00836.

6 CONFLICTS OF INTEREST

The authors declare that they have no known conflicts of interest.

7 REFERENCES

1. W. J. Price, *Spectrographic analysis of complex oxides with particular reference to slags and ores*, Spectrochimica Acta, 1953.
2. F. J. M. Maessen and P. W. J. Boumans, *Spectrochim Acta B*, 1968, **B 23**, 739.
3. I. Croudace, P. Warwick, R. N. Taylor and S. Dee, *Anal Chim Acta*, 1998, **371**, 217.
4. U. Nygren, I. Rodushkin, C. Nilsson and D. C. Baxter, *J. Anal. At. Spectrom.*, 2003, **18**, 1426.
5. S. Röllin, H. Sahli, R. Holzer, M. Astner and M. Burger, *Appl. Radiat. Isot.*, 2009, **67**, 821.
6. A. Gagne, J. Surette, S. Kramer-Tremblay, X. X. Dai, C. Didychuk and D. Lariviere, *J. Radioanal. Nucl. Chem.*, 2013, **295**, 477.
7. M. Trdin, M. Necemer and L. Benedik, *Anal. Chem.*, 2017, **89**, 3169.
8. M. Jäggi, M. Rüthi and E. J., *J. Radioanal. Nucl. Chem.*, 2019, **322**, 1279.
9. L. Benedik, A. M. Pilar and H. Prosen, *J. Environ. Radioact.*, 2020, **213**, 106144.
10. P. E. Warwick, I. W. Croudace and J. S. Oh, *Anal. Chem.*, 2001, **73**, 3410.
11. E. Braysher, B. Russell, S. Woods, M. Garcia-Miranda, P. Ivanov, B. Bouchard and D. Read, *J. Radioanal. Nucl. Chem.*, 2019, **321**, 183.
12. F. Claisse and J. Blanchette, *Physics and Chemistry of Borate Fusion*, Katanax, Quebec edn., 2018.
13. J. J. Wang, I. J. Chen and J. H. Chiu, *Appl. Radiat. Isot.*, 2004, **61**, 299.
14. N. Vajda and C. K. Kim, *Appl. Radiat. Isot.*, 2010, **68**, 2306.
15. E. P. Horwitz, R. Chiarizia and M. L. Dietz, *Solvent Extr Ion Exc*, 1992, **10**, 313.

16. E. P. Horwitz, D. R. McAlister, A. H. Bond and R. E. Barrans, *Solvent Extr Ion Exc*, 2005, **23**, 319.
17. X. X. Dai and S. Kramer-Tremblay, *Anal. Chem.*, 2014, **86**, 5441.
18. F. Claisse, in *Comprehensive Analytical Chemistry XLI*, eds. Z. Mester and R. E. Sturgeon, Elsevier, Amsterdam, 2003.
19. S. Bajo and J. Eikenberg, *J. Radioanal. Nucl. Chem.*, 1999, **242**, 745.
20. H. Bennett and G. J. Oliver, *Analyst*, 1976, **101**, 803.
21. A. Michaud and D. Larivière, Cleared, 88PA, Case # 2016-4620, 19 Sep 2016, 2016.

8 APPENDIX

Elemental composition of the inactive barite concrete as determined by ICP-OES. The inactive barite concrete was dry-ashed and fused, Si was flocculated as SiO_2 and filtrated from the solution to be measured.

Element	Inactive barite concrete [mg/g]	Detection limit [mg/g]
Al	3.0585	0.0025
As	0.0499	0.0250
Ba	3.5480	0.0004
Be	0.0012	0.0002
Ca	76.4289	0.0001
Cr	0.0075	0.0008
Cu	0.0050	0.0017
Er	0.0008	0.0008
Fe	21.0129	0.0017
Gd	0.0033	0.0008
K	1.3229	0.0083
La	0.0025	0.0008
Mg	0.7164	0.0000
Mn	0.0494	0.0004
Na	0.1581	0.0083
P	0.0416	0.0250
Pb	0.0333	0.0083
S	4.4098	0.1664
Sc	0.0009	0.0001
Si	0.2579	0.0166
Sr	6.1736	0.0001
Ti	0.1830	0.0008
V	0.0058	0.0017
Y	0.0134	0.0001
Zn	0.0383	0.0017
Zr	0.0100	0.0017

CHAPTER 10 DEVELOPMENT OF MICROFLUIDIC SYSTEMS FOR ACTINIDE SEPARATION USING FUNCTIONALISED METHACRYLATE MONOLITHS

Shuang Yu Han[a*], Bernard Treves Brown[a], Matthew Higginson[b], Philip Kaye[b], Clint Sharrad[a] and Scott Heath[C]

[a]Department of Chemical Engineering, The University of Manchester, Oxford Road, Manchester M13 9PL, UK, [b]AWE, Aldermaston, RG7 4PR, UK, [C]Department of Mechanical, Aerospace & Civil Engineering, The University of Manchester, Oxford Road, Manchester M13 9PL, UK

*Shuangyu.han@manchester.ac.uk

1 INTRODUCTION

Identification, and quantification of major actinides such as uranium (U) and plutonium (Pu), potentially associated trace elements and fission products through an intricate system of analytical chemistry are essential for quality control, environmental monitoring, and nuclear material accountancy throughout the nuclear fuel cycle.[1-3] From the front–end, ore discovery, fuel fabrication to the back–end, fuel processing, reprocessing and end of life treatment, understanding of the nature and composition of the nuclear materials are essential to ensure safe and efficient operation.[1,2]

However, nuclear materials are diverse depending on their origin, potentially containing numerous transition metals and diverse categories of fission products – from transuranic (Am, Np, Cm etc.) to lanthanides (La, Ce, Pr etc.) elements that are generated during fission reaction which may interfere with available analytical techniques.[4]

Radioactive elements such as U and Pu possess rich and overwhelming optical emissions spectrums which may amplify, distort, and interfere with optical spectrometric analytical methods such as inductively coupled plasma optical emission spectroscopy (ICP-OES) and direct current arc (DC Arc) emission spectroscopy.

Mass spectrometry based analytical systems are not significantly affected by optical interferences. However, these measurements are affected by isobaric and polyatomic interferences, where the presence of elemental isotopes or combinations of isotopes that share an identical molecular mass distort the measurement of the target analyte.[5]

To overcome the various possible interference sources, a separation or selective removal stage following sample dissolution are often implemented, either removing the target analyte or removing strongly interfering elements by solvent extraction or ion exchange chromatography.[6] However, both processes require intensive manual work are difficult to automate within glove boxes, coupled with generation of radioactive liquid, organic and solid waste in the tens of millilitre range.[1-3]

Miniaturisation of the radiochemical separation and potentially the analytical stage may be possible with microfluidics – reducing the amount of fluid manipulated to the microlitre scale. This offers an alternative analytical pathway which satisfies the ALARA (As Low as Reasonably Achievable) approach to the reduction to the usage of dangerous chemicals, samples, and generation of hazardous waste.[7] A microsystem also offers opportunities for scaled up operation through parallel processing and potential for automation.[2,3 6,8,9]

Miniaturisation of radiochemical analyses have been explored in several directions, miniaturised liquid – liquid extractions (LLE), and solid – phase extractions (SPE).[2] SPE using particulate resins are an attractive option, due to the availability of well tested and reliable commercial ion exchange or extraction resins which can be packed within a micro-channel by the means of a frit or a weir.[3,10]

However, reproducible packing of particulate resins into micro-channels remains difficult, as is consistent resin immobilisation onto the microfluidic device.[2,6] For certain applications, functionalised resins at this small scale may not be commercially available and may have to be produced in small batches, which generates additional waste and reproducibility issues.

A second option is based on the application of polymer monoliths that can be fabricated within the micro-channels directly through either exposure to ultra-violet light or elevated temperature.[6,12,13]

A typical polymer monolith is a piece of continuous solid hydrophobic porous structure produced from a mixture of polymer monomers, free radical initiators, and solvents. (See Figure 1).

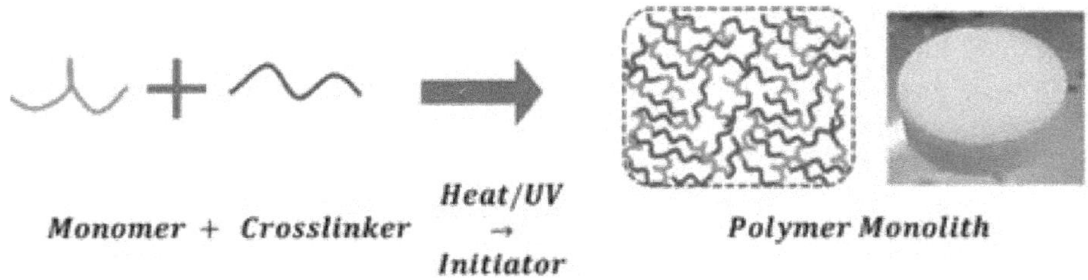

Figure 1. *Synthesis of a typical monolithic material, functional monomers and the crosslinker polymer forms a close network with the assistance of a free radical initiator, presence of solvents control the level of porosity within the porous material.*

The monoliths, typically possessing a large amount of macro and mesopores, were originally developed for separation of biological materials and has since extended their application to solid support for extraction chromatography and catalysts.

Bruchet et al.[13] and Losno et al.[6] constituted the only examples of radiochemical separation of nuclear samples with a functionalised monolith as the extraction material. Bruchet et al. preferred covalent bonding of extractant directly onto the monolith fabricated with glycidyl methacrylate (GMA), while Losno et al. impregnated several organic extractants directly onto a polymer monolith surface fabricated with allyl methacrylate, using methods similar to the manufacturing of many commercially available extraction resins.[14]

This work aims to highlight the potential application of solvent impregnated methacrylate monoliths as solid phase extraction materials for radiochemical analysis in acidic conditions. Several methacrylate monomers – allyl methacrylate (AMA), methyl methacrylate (MMA) and butyl methacrylate (BMA) are produced in situ within syringes of different dimensions and impregnated with tributyl phosphate (TBP).

The monolith performances are then evaluated with U retention and extraction with concentrated and dilute nitric acid. With the aim of integrating functionalised polymer monoliths for application within a microfluidic system for actinide separation, assay and trace element analysis.

2 EXPERIMENTAL

2.1 Materials

BMA, MMA, Ethylene glycol dimethylacrylate (EDMA), 2,2-Dimethoxy-2-phenylacetophenone (DMPA), propan-1-ol, TBP and 1,4-butanediol were purchased from Sigma-Aldrich (Gillingham, UK). AMA were purchased from Tokyo Chemical Industry Ltd. (Oxford, UK). Nitric acid (Primar Plus™) was purchased from Fisher Scientific (Loughborough, UK). Calibrated Uranium solutions used were provided from stocks produced at the University of Manchester.

2.2 Monolith preparation methodology

The procedures for monolith synthesis were adapted from Losno et al.[6]. The polymerisation ratio of 1 wt% water, 24 wt% 1,4-butanediol, 35 wt% 1-propanol, 24 wt% EDMA, 16 wt% functional monomer, 1 wt% DMPA was used.

The polymerisation mixture was filled to approximately 2 ml in the 6 ml polypropylene (PP) syringe, and approximately 0.4 ml in the 1 ml polyethylene (PE) syringe. The synthesis of monolithic phase was achieved by exposure to a 400 W UV flood lamp emitting primarily UVA wavelength for 10 minutes.

The formed monolith is then flushed with methanol to remove any unreacted polymer, dried over 24 hours and weighed. Impregnation of the monolith was achieved by flowing undiluted TBP through the monolith. The columns were then rinsed with water until no solvent spots can be seen in the eluted liquid and weighed again.

2.3 Characterisation

Scanning electron microscopy imagining was completed with a Quanta 200 scanning electron microscope after metal spin coating.

Measurement of the U concentration in sample and eluted solutions was achieved with a high-resolution energy dispersive X-ray fluorescence spectrometer (Rigaku NEX DE EDXRF), calibrated with U standard solution provided by the University of Manchester

For determination of monolith retention characteristics, a distribution coefficient (Dw) was used to gauge U retention performance, the columns are conditioned with 1 M, 4 M and 8 M nitric acid for 20 minutes.[6] A solution of 1000 ppm U in nitric acid mixture were passed through the columns and the distribution coefficients were calculated as:

$$Dw = \frac{C_0 - C_F}{C_F} \times \frac{V}{Monolith\ Mass} \tag{1}$$

Where Dw is the distribution coefficient of the element, C_0 is the initial U concentration in the aqueous phase, C_F is the final U concentration in the aqueous phase, V is the volume of the sample phase, and mass of the monolith phase in the syringe.

3 RESULTS AND DISCUSSION

Synthesis and impregnation of the monolithic materials were mostly successful with all the monomers used in this study. However, breakoffs of some monolith pieces were noted during the initial methanol wash, possibly as a side effect of removing unreacted polymers and solvents.

Distribution coefficients are a useful metric in gauging the retention performances of the extractant by considering the fraction of the analytes that are removed, volume of the sample and mass of the extraction medium. As a general rule and as demonstrated by the batch absorbance experiments (Figure 2), a Dw>100 demonstrated a high U retention by the monolith, while Dw<100 generally indicates incomplete or relatively weak U retention.

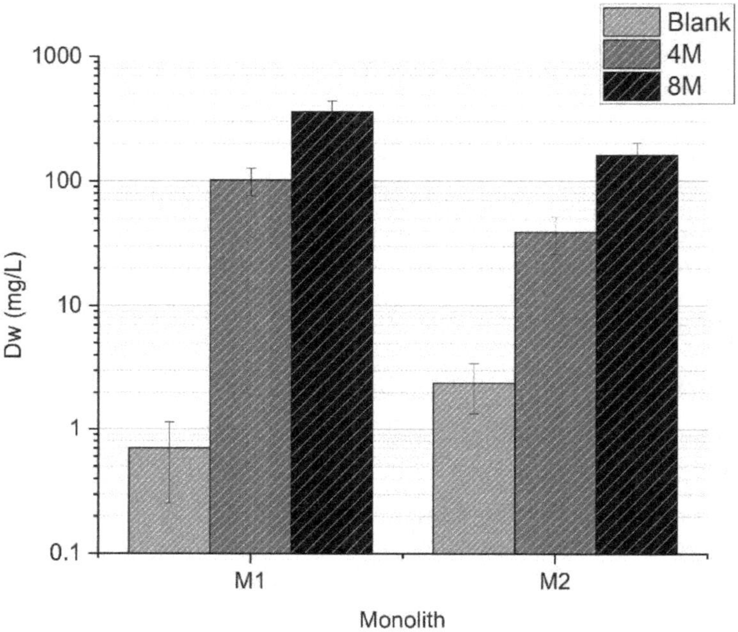

Figure 2. *Batch uranium extraction experiments with virgin and TBP impregnated AMA monoliths, assessing for U uptake in a macro environment.*

In this work, unimpregnated AMA monoliths demonstrated little to no retention of U, enabling identification and isolation of the effect of TBP impregnation. The retention characteristic of impregnated monoliths is represented as lines in figure 2 as the nitric acid concentration varies in the U sample medium.

In general, it appears that both types of monomers used, and the concentration of nitric acid used have significant impact on the final U retention. Higher U retention with commercially available TBP resins occurs at higher nitric acid concentrations, while an alternative behaviour is observed in this work, depending on the type of the monoliths, increasing nitric acid concentration may have negative effect on the overall U recovery.[14]

Impregnated MMA demonstrated the highest U retention behaviour out of all 2 ml size monoliths, which are also in line with the commercial resin's behaviour.[14] The ten-fold Dw value increase with increasing nitric acid concentration (1 to 4 to 8 M), achieving the highest degree of U retention at the highest nitric acid concentration at 518 (±260), is comparable to commercial TBP resins at these concentrations.

An increasing retention trend with nitric acid is also observed in the AMA-TBP monolith. However, overall retention of U at lower acid concentration is almost non-existent. At the highest nitric acid concentrations used, Dw is barely above 2, indicating no significant U uptake regardless of increasing acid level.

For BMA-TBP monoliths, Dw reached its peak at 4 M nitric acid and reduced to 26% of its peak values when exposed to 8 M nitric acid samples.

There are limited studies completed on the removal of actinides with solvent impregnated polymer monoliths. However, the ineffectiveness of TBP-AMA monoliths did not fit with the results obtained by Losno et al.

Several possibilities exist, the walls of the polymer syringes were not functionalised compared to previous work, causing the monolith produced to not be chemically attached to the channel walls. This may have led to channelling near the wall as the geometry of the monoliths are interrupted, which is a problem that has been identified in conventional packing in separation industries.[15]

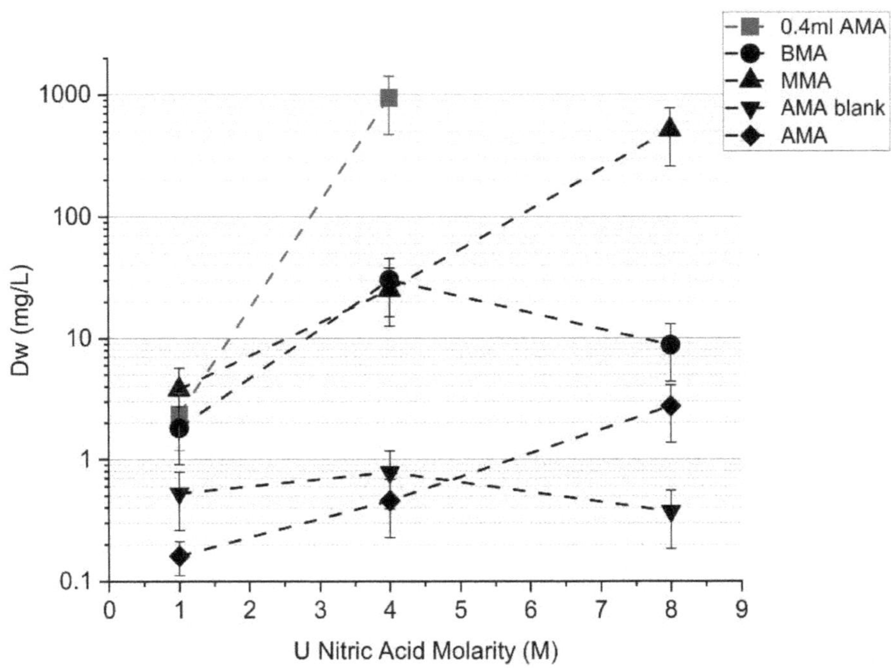

Figure 3. *Calculated U distribution coefficients for several TBP impregnated monoliths produced from different monomers. BMA, MMA, AMA. A blank monolith made from AMA and a 0.4 ml AMA monolith fabricated in 1 ml PE syringes are included.*

In an effort to investigate further, AMA-TBP monoliths were produced again within a PE syringe with a much smaller diameter and a greater length. As shown in Figure 3. This system has managed to reach significantly higher Dw at 950 (±475) when compared to AMA-TBP monolith produced in the 6 ml PP syringe. This value is also in line with the Dw as determined by Losno et al. in their work, demonstrating the potential benefits of size reduction and conversion to microfluidic devices that may prove beneficial to the performances of these impregnated monolithic structures.

It is important to note that several potential factors could be responsible for the increase in U retention effectiveness as the scale decrease. The smaller syringes used are manufactured from PE instead of PC, and its surface chemistry may have aided either monolith formation or a more closely linked structure near the wall.[16]The increased volume to length ratio may have also aided in U uptake despite comparable flowrate ratio to the larger monoliths. Further testing with functionalisation and attachment of monolithic structures to the channel walls also need to be carried out to isolate the impact of potential channelling that may occur in the system.

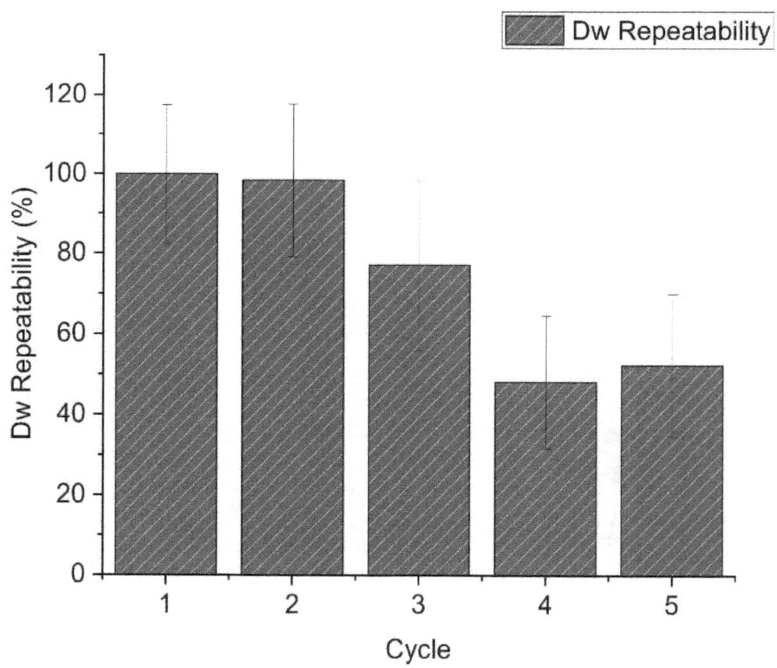

Figure 4 *Calculated Dw comparisons after five repeats for U extraction were carried out at 4 M with TBP impregnated MMA.*

In order for the monolith produced to be truly used as a replacement for commercial resins, reusability must be considered to determine when the monolithic phase should be replaced after certain amount of uses. Five repeated runs of U retention and elution were carried out using MMA-TBP monolith with 4 M nitric acid sample five times and the resulting Dw values are shown in Figure 4.

The results indicates that while impregnated MMA-TBP monoliths are stable in the first two cycles, it suffers from significant deterioration of retention performances over prolonged period, reducing the U Dw value by 50%, from approximately 60 to 30 in five repeated runs.

4 CONCLUSION

In this work, a polymer monolith made from copolymerisation of functional monomers AMA, BMA and MMA with crosslinker EDMA was prepared and functionalised with an organic extractant within PC syringes. All three types of TBP impregnated monoliths have demonstrated successful retention of U to varying degrees. Out of all 2 ml sized impregnated monoliths produced, MMA – TBP monolith demonstrated comparable Dw at 25 (±12) and 518 (±260) when compared to commercial TBP™ resins at 4 and 8 M nitric acid concentrations respectively.

Improved extraction efficiencies were further observed with AMA-TBP monolith synthesised within a smaller PE syringe, reduction in size may lead to better performances due to elimination of channelling or the increased diameter to length ratio, demonstrating the potential for performance increase in small scale devices. However, further characterisation and experimentations at this scale is required.

While the selected impregnated monoliths showed good U retention characteristics and can be reused a few times, the overall performances will deteriorate significantly after two or three cycles of retention and elution. Attempts at further miniaturisation and implementation of impregnated methacrylate monolith in a microfluidic device is already under investigation. Further work into improvement of reusability or online re-functionalisation of impregnated monoliths with additional extractants needs to be investigated in order for realistic application of these monolithic extraction phases.

5 CONFLICT OF INTEREST

The authors declare that there is no conflict of interest.

6 ACKNOWLEDGEMENTS

This work is funded and supported by Industrial Cooperative Awards in Science & Technology (CASE) received through Engineering and Physical Sciences Research Council (EPSRC) and AWE.

7 REFERENCES

1. C. Mahan, S. Bonchin, D. Figg, D. Gerth and C. Collier, *Journal of Analytical Atomic Spectrometry,* 2000, **15**, 929.

2. J. Gao, B. Manard, A. Castro, D. Montoya, N. Xu and R. Chamberlin, *Talanta,* 2017, 167, 8.
3. C. Mariet, A. Vansteene, M. Losno, J. Pellé, JP. Jasmin, A. Bruchet, G. Hellé, *Micro and Nano Engineering*, 2019, **3**, 7.
4. B. Manard, S. Metzger, S. Wysor, V. Bradley, B. Roach, N. Alex Zirakparvar, K. Rogers, D. Bostick, B. Ticknor and C. Hexel, *Applied Spectrometry*, 2021, **75**, 556.
5. E. Wylie, B. Manard, C. Quarles, L. Meyers and N. Xu, *Talanta*, 2018, **189**, 24.
6. M. Losno, J. Pellé, M. Marie, I. Ferrante, R. Brennetot, S. Descroix and C. Mariet, *Talanta*, 2018, **185**, 586.
7. M. Higginson, K. Palmer, J. King, B. Dawkins, T. Huggins, L. Ingman, F. Taylor, N. Xu and P. Kaye, *Journal of Radioanalytical and Nuclear Chemistry*, 2019, **320**, 689.
8. H. Hotokezaka, M. Tokeshi, M. Harada, T. Kitamori and Y. Ikeda, *Progress in Nuclear Energy*, 2005, **47**, 1.
9. J. Zhang, Z. Qin, D. Deng, J. Liao, X. Wei and N. Zhang, *Journal of Analytical Atomic Spectrometry*, 2016, **31**, 934.
10. A. Bickham, C. Pang, B. George, D. Topham, J. Nielsen, G. Nordin and A. Woolley, *Analytical Chemistry*, 2020, **92**, 12322
11. D. Rahmi, Y. Takasaki, Y. Zhu, H. Kobayashi, S. Konagaya, H. Haraguchi and T. Umemura, *Talanta*, 2010, **81**, 1438.
12. M. Losno, I. Ferrante, R. Brennetot, J. Varlet, C. Blanc, B. Grenut, E. Amblard, S. Descroix and C. Mariet, *Micromachines*, 2016, **7**, 45.
13. A. Bruchet, V. Taniga, S. Descroix, L. Malaquin, F. Goutelard and C Mariet, *Talanta*, 2013, **116**, 488.
14. C. Dirks, N. Vajda, E. Kovács-Széles, A. Bombard. S Happel. Characterization of a TBP Resin and development of methods for the separation of actinides and the purification of Sn" Poster presented at the 17th Radiochemistry conference, Mariánské Lázně (Czech Republic), 11 - 16 May 2014
15. S. White and C. Tien, *Warme- und Stoffubertragung*, 1986, **21**, 291.
16. R. Groarke and D. Brabazon. *Materials (Basel)*, 2016, **9**, 446.

CHAPTER 11 RESEARCH REACTOR SUPPORT FOR NUCLEAR FORENSICS STUDIES AND THE DEVELOPMENT OF A COMPANION GRADUATE COURSE

S. Landsberger[a]* and D. Haas[a]

[a]University of Texas at Austin, Nuclear Engineering Teaching Lab, Pickle Research Campus, R-9000, 10,000 Burnet Road, Building #159, Austin, Texas, USA, 78712

*s.landsberger@mail.utexas.edu

1 INTRODUCTION

In the 15th century the Latin term "forensis" came to being meaning to belong to the public forum. Later the word forensics was incorporated into the English language to mean, "pertaining to, connected with, or used in courts of law or public discussion and debate". In recent years, the word forensics has been popularized in the media and pop-culture in the many television shows and movies as part of crime scene investigations. The growth of adaptive tools in nuclear forensics has dramatically increased in the last two decades encompassing a wide range of technologies to be used in the investigation and analysis of nuclear materials before they are used in an attack, the examination of radioactive debris following a nuclear event or the study of the pedigree or provenance of nuclear materials in non-proliferation. However, analytical techniques in nuclear forensics are also used in radioecology to ascertain environmental radioactivity that deal with naturally occurring radioactive materials (NORM) or technologically enhanced NORM, commonly known as TENORM. As well, nuclear forensics has been successfully integrated to study the movement of plutonium and fission products from the weapons tests[1,2] and the two infamous nuclear reactor accidents of Chernobyl and Fukushima.[3,4]

Lastly, nuclear forensics was used to determine the cause of the assassination death of former KGB operative Litvinenko[5] by ^{210}Po and possibly the same for Chairman Yasser Arafat[6] of the Palestine Liberation Organization. An overview of the various types of nuclear forensics research is depicted in Figure 1.

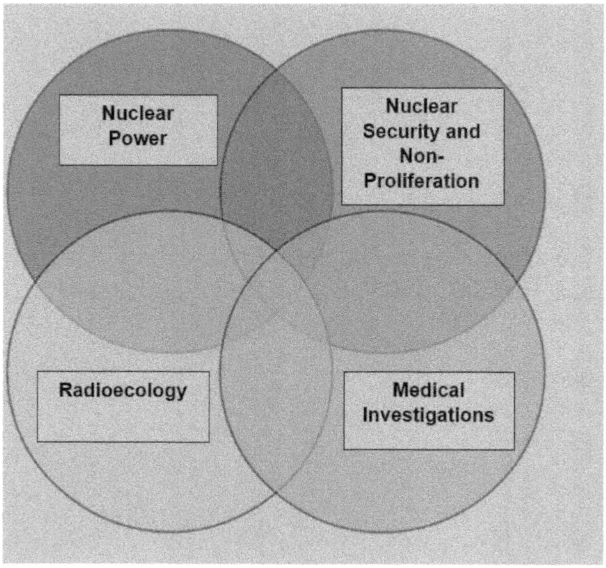

Figure 1 *Overview of Differing Nuclear Forensics Research.*

In the past twenty years, the Nuclear and Radiation Engineering program has been at the vanguard of developing experimental facilities and methodologies in nuclear forensics at the TRIGA 1.1 MW research reactor, mainly in post detonation related research, but also in NORM and TENORM investigations, and radiation dose assessments. More recently, we have developed a graduate level course in nuclear forensics with a set of related laboratories and assignments and several invited guest speakers covering a wide area crosscutting relevant technologies. The intent of this chapter is to highlight the experimental and educational achievements of our nuclear forensics program which has been supported by Department of Energy, Department of Homeland Security, Los Alamos National Laboratory, Oak Ridge National Laboratory, Pacific Northwest National Laboratory, and North Carolina State University. Throughout our program, multiple MS and PhD graduate students who have been financially supported have gone to find permanent positions in these national laboratories and government agencies.

2 NUCLEAR FORENSICS RESEARCH

The ability to have access to beams of neutrons and produce fission products is a huge advantage when undertaking nuclear forensics research. The 1.1 MW

TRIGA reactor that is housed in the Nuclear Engineering Teaching Lab, has been effectively used in nuclear forensics research. There is no intent to explain the in-depth research results, but rather emphasize that all these publications have uniquely resulted by the use of a research reactor. Below is an abbreviated list of several recent publications in the areas of nuclear forensics described in the introduction:

1. A cryogenic gas irradiation facility in the 1.1 MW TRIGA reactor was designed to produce radioactive xenon and argon for environmental studies.[7]

2. The gaseous diffusions of neutron activated krypton and xenon were investigated in silicate and sodium chloride media at multiple pressures.[8]

3. Implications of conflicting cross sections for the $^{40}Ca(n,\alpha)^{37}Ar$ reaction as applied to nuclear explosion monitoring.[9]

4. A method for measuring radioxenon emanation with cryotrapping and γ spectroscopy.[10]

5. Evaluation of independent and cumulative fission product yields with gamma spectrometry.[11]

6. Elemental analysis of the geothermic microecology of the Los Azufres (Mexico) volcanic complex.[12]

7. Investigation of bauxite residue (red mud) in terms of its environmental risk.[13]

8. Investigation of Jordanian uranium resources in carbonate rock.[14]

9. Non-destructive determination of ^{235}U, ^{238}U, ^{232}Th and ^{40}K concentrations in various consumed nuts and their implication on radiation dose levels to the human body.[15]

10. A feasibility study on the determination of ^{90}Sr food-chain transfer using stable strontium as a surrogate and neutron activation analysis.[16]

11. Disequilibrium in the uranium and actinium series in oil scale samples.[17]

3 NUCLEAR FORENSICS COURSE

There are currently several books, papers and review articles dealing with nuclear forensics that cover a broad range of topics.[18-28] The intent of this one-month intensive course was to cover the basic radioactivity topics including alpha and beta decay, various equilibria, interactions of radiation with matter including alpha and beta particles, photons and neutrons, and a real understanding of decay schemes (an aspect overlooked in many nuclear science and engineering courses), concept of a nuclear reactor, gamma-ray self-attenuation, and counting statistics and detection limits. Besides these usual lectures on the basics of nuclear phenomena, a successful effort was made to include historical background. One lecture entitled "Pop Culture Goes Nuclear" (Mihir Pai, Graduate Student) details the decades of change of how nuclear weapons were viewed by the media with many video clips of classic Hollywood and some British cartoons, movies and documentaries. Another lecture provides detailed background on the "Making of the Atomic Bomb" (Richard Brooks, Consultant). From a pedagogical view, it was rewarding to the students to invite other subject matter experts to present lectures that improved the quality and breadth of the course. History and Effects of Radiological Weapons and Radiation Portal Monitors were given by William Charlton, while Chemistry of the Actinides and Fission Products, Chemical Equilibria, Chemical Separations Chemical Separations, Nuclear Fuel Cycle, Nuclear Fuel Reprocessing, were given by Joseph Lapka, both staff members at the Nuclear Engineering Teaching Laboratory. Robert Anderson (Nuclear and Applied Robotics) gave a lecture in Application of Robotics Locating Radioactive Sources with Gamma Ray Shielding. The invitation of professors from other universities and research scientists from US National Laboratories made the course more relevant, interesting, and well-appreciated by the students. These lectures included the following: Post Detonation Forensic Analysis of Nuclear Fallout Using On Site Gamma Spectroscopy (John Mattingly, North Carolina State University); Measurement of Uranium Isotope Ratios in Keratinous Materials (John Brockman, University of Missouri); Materials Analysis for Nuclear Forensics (Roger Kapsimalis, Oak Ridge National Lab); Solid-state Uranium Chemistry in the Nuclear Fuel Cycle (Tyler Spano, Oak Ridge National Lab); Various Applications of Machine Learning/Data Analytics for Non-proliferation (Ken Dayman, Oak Ridge National Lab); Passive and Active Neutron Non-Destructive Assay Techniques (Cole Thompson, Los Alamos National Laboratory); New Scintillators and their Integration into Detector Systems (Nerine Cherepy, Lawrence Livermore National Laboratory); Cat Litter, Nitric Acid, Plutonium, and $2.4 Billion: Pacific Northwest National Laboratory Role in the Investigation of 2014 Contamination at the Waste Isolation Pilot Plant (Jon Schwantes, Pacific Northwest National Laboratory); and Development of Fieldable Alpha Spectrometry Systems for Non-proliferation and Nuclear Forensics; (David Chichester, Idaho National Laboratory).

3.1 Homework assignments

Five homework assignments were given in the traditional way with problem solving for activity measurements, secular equilibrium and simple shielding calculations of Pu and U sources. In the first assignment, the students needed to express their opinion on the ethical use of nuclear weapons used in Hiroshima and Nagasaki and their use in the future. As well, each assignment had additional questions on the invited presentations. This ensured that the students did not just listen to the presentation, but actually deciphered its contents to be able to answer questions. Since all the lectures were recorded by ZOOM, students could easily revisit the lectures.

3.2 Laboratories

In general, there is a wide variety of laboratories that can be performed in nuclear forensics. These experiments can be simple to complicated ones in radiochemistry, alpha spectrometry, gamma-ray spectrometry, liquid scintillation, inductively coupled plasma-mass spectrometry, thermal ion mass-spectrometry, etc. Undoubtedly, it is impossible to implement all these experiments in one laboratory section in a nuclear forensics course. Because of our expertise in gamma-ray spectrometry a series of experiments were developed around this technique. These laboratories included half-life measurement and secular equilibrium using a ^{137}Cs-^{137m}Ba generator, counting statistics, shielding of beta particles and gamma rays, basics of gamma ray spectrometry, gamma-ray self-attenuation using high-Z material in oil scale, and fission product identification using induced naturally occurring uranium irradiated with neutrons from our 1.1 MW TRIGA research reactor. This latter experiment included construction of an efficiency curve, counting statistics, detection limits and detailed usage of decay schemes. Because of COVID-19 restrictions and our on-line distance-learning program, an overview of the experiments were also recorded on a smart phone and videos and data were uploaded on to the educational site Canvas.

3.3 Final Examinations

The final examination was typically a take-home exam where the students needed to critique some nuclear forensics /nuclear non-proliferation/nuclear security event. Examples include:

Give a complete historical description/assessment of the treaties between the former USSR and the USA in limiting the production of nuclear weapons.

Give a complete historical description/assessment of the history of the North Korean nuclear weapons program.

Give a complete historical description/assessment of the history of the Iranian nuclear weapons program.

Discuss the air crash in 1992 when an Israeli El-Al Cargo 747 Airplane travelling from JFK in New York to Ben Gurion Airport in Israel lost one of its engines and slammed into an apartment building in Amsterdam and 150 kg of depleted uranium was used as a balance weight for the aircraft.

Give a complete historical description/assessment of the history of the Indian and Pakistani nuclear weapons programs.

What technologies are expected to be implemented in the GEN-1V reactors advanced reactors to make them more nuclear proliferation resistant?

Discuss the possible death of Chairman Yasser Arafat of the PLO by ^{210}Po.

4 CONCLUSIONS

Nuclear forensics research now encompasses a very wide area of research and development in instrumentation, safeguards, drones and robotics, chronometry, machine learning language, etc. Research reactors can play a critical role in providing experiments with production of isotopes, fission products and neutron beams to support nuclear forensics. Graduate course development with and without laboratories on a particular subject can vary for each academic institution. Such technological advances as mentioned above, cannot be continuously permeated into courses. However, inviting subject matter experts clearly adds value to the students. We have developed a template in theory and practical hands-on experiments. Covid-19 has also fine-tuned our capabilities of offering on-line laboratories to distance learning students.

5 ACKNOWLEDGEMENTS

The authors acknowledge the support of the Department of Energy Radiochemistry Education Award Program, Los Alamos National Laboratory, Oak Ridge National Laboratory, Pacific Northwest National Laboratory, Department of Homeland Security (DHS) and DHS Nuclear Forensics Education Award Program for the past two decades for financial support for various research projects and multiple graduate students.

6 REFERENCES

1. D. Feng, et al., Journal of Environmental Radioactivity, 2021, 242, 106792.

2. V. Drozdovitch, F. Vathaire and A. Bouville, Journal of Environmental Radioactivity, 2020, 214–215,106176

3. V. Silliková, J. Jakubčinová, M. Horník, I. Gomola, S. Dulanská, Journal of Radioanalytical and Nuclear Chemistry, 2022, 331, 1275.

4. M. Yama and S. Oikawa, Journal of Radioanalytical and Nuclear Chemistry, 2022, 331, 2689.

5. R. B. McFee and J. B. Leikin, Seminars in Diagnostic Pathology, 2009, 26, 61.

6. P. Froidevaux, et al., Forensic Science International, 2016, 259, 1.

7. C. C. Hudson, R. Lester, T. Tipping and D. Haas, Journal of Radioanalytical and Nuclear Chemistry, 2022, https://doi.org/10.1007/s10967-022-08515-7.

8. J. L. Lapka, D. A. Haas and Justin D. Lowrey, Journal of Radioanalytical and Nuclear Chemistry, 2022, https://doi.org/10.1007/s10967-022-08533-5

9. K. A. Khiloni, B. A. de Luna and D. Haas, Journal of Radioanalytical and Nuclear Chemistry, 2022, https://doi.org/10.1007/s10967-022-08522-8.

10. M. e Blood, S. R. Biegalski, D. A. Haas, E. J. Artnak,, Journal of Radioanalytical and Nuclear Chemistry, 2018, 318, 305.

11. K. Dayman, S. Biegalski, D. Haas A. Prinke, S. Stave, Journal of Radioanalytical and Nuclear Chemistry ,2016, 307, 2239. DOI 10.1007/s10967-015-4491-9.

12. W. Abuhani, et al., Journal of Radioanalytical and Nuclear Chemistry, 2019, 322, 1365.

13. B. Ozden, C. Brennan and S. Landsberger, Journal of Radioanalytical and Nuclear Chemistry, 2019, 319, 339.

14. A. Alsabbagh, L. Zaidan, I. Harahsheh, N. Ba Sunbul and S. Landsberger, Journal of Radioanalytical and Nuclear Chemistry, 2016, 308, 1063.

15. S. Landsberger, D. Tamalis and R. Lara, Journal of Radioanalytical and Nuclear Chemistry, 2016, 307,1065.

16. C. Brennan et al., Journal of Environmental Radioactivity, 2019, 105988.

17. S. Landsberger, D. Tamalis, C. Leblanc and M. D. Yoho, Journal of Environmental Radioactivity, 2017, 166, 126.

18. K. J. Moody, P. M. Grant and I. D. Hutcheon, Nuclear Forensics Analysis, CRC Press, Boca Raton, Florida, USA, 2nd edition, 2015.

19. V. Fedchenko, The New Nuclear Forensics, Analysis of Nuclear Materials for Security Purposes, V. Fedchenko, Oxford University Press, Oxford UK, 2015.

20. G. W. Philips, D. Nagel and T. Coffey, A Primer of the Detection on Nuclear and Radiological Weapons, Center for Technology and National Security Policy, National Defense University 2005. http://large.stanford.edu/courses/2018/ph241/imanbayev2/docs/phillips.pdf (October, 2022).

21. International Atomic Energy Agency, IAEA-TECDOC-1896, 2018. https://www.iaea.org/publications/13648/nuclear-forensics-beyond-the-science (October 2022).

22. International Atomic Energy Agency, Development of a National Nuclear Forensics Library: A System for the Identification of Nuclear or Other

Radioactive Material out of Regulatory Control, IAEA-TDL-0092018, 2018, https://www.iaea.org/publications/13438/development-of-a-national-nuclear-forensics-library-a-system-for-the-identification-of-nuclear-or-other-radioactive-material-out-of-regulatory-control (October 2022)

23. International Atomic Energy Agency, Identification of High Confidence Nuclear Forensics Signatures, IAEA-TECDOC-1820, 2017. https://www.iaea.org/publications/12231/identification-of-high-confidence-nuclear-forensics-signatures (October 2022)

24. International Atomic Energy Agency, Nuclear Forensics in Support of Investigations, STI/PUB/1687, 2015, https://www.iaea.org/publications/10797/nuclear-forensics-in-support-of-investigations (October 2022)

25. International Atomic Energy Agency, Advances in Nuclear Forensics: Countering the Evolving Threat of Nuclear and Other Radioactive Material out of Regulatory Control, STI/PUB/1706, 2015, https://www-pub.iaea.org/MTCD/Publications/PDF/Pub1706_web.pdf, (October, 2022).

26. Z. Karpas, Analytical Chemistry of Uranium: Environmental, Forensic, Nuclear, and Toxicological Applications, CRC Press, CRC Press, Boca Raton, Florida, USA, 2014.

27. F. E. Stanley, A. M Stalcup and H. B. Spitz, Journal of Radioanalytical and Chemistry, 2013, 295, 1385.

28. F. Taylor, M. Higginson, O. Marsden and J. Schwantes, Journal of Radioanalytical and Chemistry, 2020, 323, 415.

CHAPTER 12 STUDIES ON DISTRIBUTION OF ^{210}Po AND ^{210}Pb ACTIVITY AND PHYSICO CHEMICAL EFFECTS OF UNDISTURBED ALLUVIAL CLAY SOIL IN BELAGAVI DISTRICT, KARNATAKA, INDIA

V. Kamalakar[a*], P. Vinutha[b], S. Achari[a] and Y. Narayana[a]

[a]Department of Physics, Mangalore University, Mangalagangothri-574 199, India, [b]Department of Physics, Bearys Institute of Technology, Mangalore-574 153, India

*narayanay@yahoo.com

1 INTRODUCTION

The major background radiation exposures among the population are naturally occurring radionuclides in soil. Geographical arrangements and rocks that are affluent in phosphates, granite and salts contain potassium, uranium and thorium and get migrated to other places through rainwater and river flows. ^{210}Po and ^{210}Pb, two naturally occurring radioelements present from the decay of ^{238}U, are considered to be major contributors to the internal radiation dose of humans.[1,2] ^{210}Po is an alpha emitter with a relatively short half-life (138 days). ^{210}Po is the only one with a half-life long enough to play a significant role in environmental processes. The major source of ^{210}Po is approximately 60% of the atmospheric ^{210}Po is originated from the atmospheric dust. The extraneous source of ^{210}Po in the atmosphere includes forest fires, fossil fuel, volcanic eruption, migrated sea salt aerosols, and resuspension of soil particles. These sources contribute less than 10% of the total fluxes of ^{210}Pb and ^{210}Po to the atmosphere. ^{210}Pb is also a natural radionuclide emitting beta and gamma radiation, with a relatively long half-life (22.3 years). Lead (Pb) is a widespread, non-biodegradable chemical contaminant found in the soil. Once it enters the food chain through plant uptake, it is considered as a serious threat to the environment. The high specific activity of ^{210}Po (1.66×10^{14} Bq/g) is a key aspect of its radiochemistry and because of this high specific activity, ^{210}Po is considered as one of the most toxic naturally occurring radionuclides.[3] A finite amount of ^{210}Po in the atmosphere is produced from the decay of ^{210}Pb (via ^{210}Bi). The second major source of ^{210}Po in the atmosphere is atmospheric dust. Approximately 60% of the atmospheric ^{210}Po is originated from the atmospheric dust.[4,5] The spatial distribution of radionuclide in soil, and related external radiation exposure to public are not uniform but depend chiefly on the geological background and soil type of each section in the world.[6] Therefore, the analysis of soil matrices for radionuclides is significant for the estimation of radiation exposure to humans and animals and as indicators for radiological contamination in environment. Measurement of terrestrial air absorbed dose rate is also crucial since gamma radiations provide information concerning excess lifetime cancer risks. NORM, the amount of ^{210}Po and ^{210}Pb has increased in the environment because of many industrial activities such as the operation of power plants burning fossil fuels, agriculture, the fertilizer industries, and exhaust fumes from vehicles.[7,8] Numerous researchers working in geochemistry, hydrogeology, and physics have studied the polonium and lead activity concentration in various media. A detailed review of literature reveals that no earlier studies has conducted in the study area on ^{210}Po and ^{210}Pb activity. Therefore, a scintillation based alpha counts detector is used for the measurement of alpha activity in the environmental soil samples. The determination of ^{210}Po and ^{210}Pb in various environmental soil samples, the correlation between the ^{210}Po and ^{210}Pb activity, and its influence of physico chemical parameter on the specification of radionuclides in the soil of Belagavi region, Karnataka, India has been studied. This is the first report, and it discusses the baseline data of ^{210}Po and ^{210}Pb in soil samples collected from the studied area.

2 MATERIAL AND METHODS

2.1 Geological survey of Belagavi district

Belagavi is a district in Karnataka, located in the state's north-western section, east of the Western Ghats. It is situated between 15° 00' and 17° 00' north latitudes and 74° 00' and 75° 00' east longitudes, with a total size of 13,444 square kilometres. It is bordered on the southwest by the state of Goa, and on the west and north by the state of Maharashtra. In general, the climate of the district can be characterised as semi-arid. Throughout the year, the average temperature ranges from 27 °C to 35.7 °C, while the minimum temperature ranges from 13.9 °C to 20.6 °C.[9,10] Summers in the district are hot and dry, with pleasant winters. From March to May, the hot season lasts, with average high temperatures of 35.7 °C. The average annual rainfall in the Belagavi district is 808 mm. Iron ore, manganese, bauxite, clay, and limestone are some of the key economic minerals found in the Belagavi district.[11] The Banded ferruginous quartzite and schist, granite and gneissic granites due to peninsular gneiss, sandstone, limestone, dolomite, Kaladgi formation, quartzite, shale and basalt (Deccan Trap), and laterite formations are among the geological formations found in Belagavi district.

2.2 Sampling method

Following a protocol, geology was used to identify sampling areas, and undisturbed soil samples were obtained from all the locations.[12] During the summer, soil samples were collected and the sampling took place on flat ground or mild slopes. During the summer season, the study area got a brief period of inconsistent rainfall, which happens every year. The grass on the upper surface had been removed. As illustrated in Figure 1, soil samples were taken from the top surface of the sampling station.

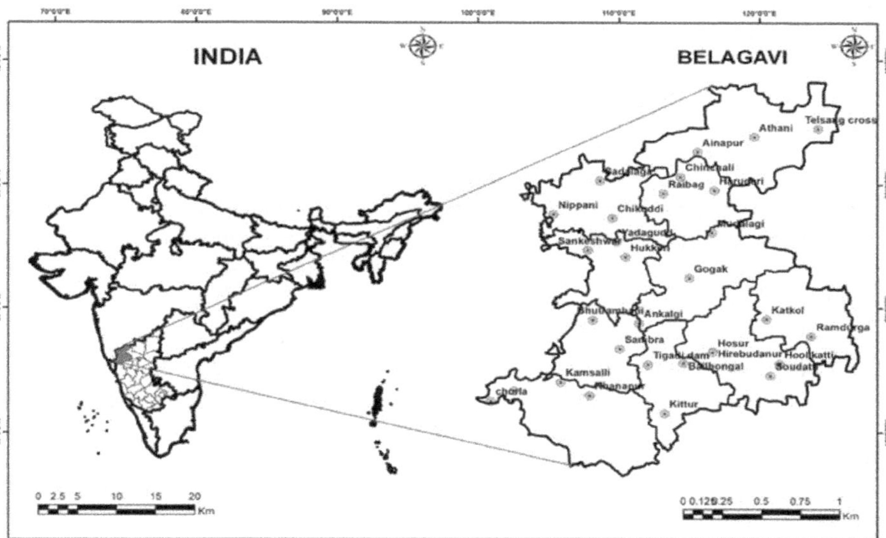

Figure 1 *Map of the Belagavi district with sampling station.*

2.3 Determination of moisture percentage

The moisture (percentage) in the soil was determined using the weight loss method. The samples were dried at 110 °C in an air-circulated oven until all moisture had evaporated, after which the moisture percentage was calculated using equation (1).

$$Moisture\ (\%) = \frac{(A - B)}{B} \times 100 \tag{1}$$

where A – Sample wet weight (grams), B – Sample dry weight (grams).

2.4 Determination of organic matter percentage

The loss of ignition method is a widely used method for determining the amount of organic matter present in soil. The oven-dried soil samples were sieved through 2000 mm and placed in a China dish to weigh for a 20 g soil sample in a chemical balance before being stored for 24 hours in a muffle furnace at 550 °C. The amount of weight lost when biological matter is present is exactly proportional to the amount of weight lost during heating. The weight loss was used to calculate the percentage of organic materials.[13]

$$Organic\ matter\ (\%) = \frac{(M_2 - M_3)}{(M_2 - M_1)} \times 100 \tag{2}$$

Where M_1 is mass of the crucible, M_2 mass of crucible plus oven-dried sample and M_3 mass of crucible plus oven-dried sample after ignition.

2.5 Determination of pH of soil

A pH tester was calibrated using a 4 and 7 pH regular buffer solution and the EUTECH (model no: 30) tester was used to determine the pH of the soil-water solution.[14] Oven dried earth was blended with distilled water in a 1:2.5 ratio. After the suspension has settled for 30 minutes, a pH metre is used to record the pH values of the soil sample, and the process is repeated for all soil samples.[15]

2.6 Determination of texture of soil

Mechanical investigation of soil samples was carried out using the international pipette system. The pipette method is based on Stoke's law, which asserts that particles in a liquid media fall or settle at different speeds. A 20 g soil sample was weighed using a chemical balance and placed in a beaker to be treated with hydrogen peroxide at 90 °C. A dispersion agent of around 10 mL was thoroughly mixed for a few minutes before being introduced to the soil-water solution (3.6 percent sodium hexametaphosphate, and 0.8 percent sodium carbonate).[16,17] For approximately 10 minutes, the solution was aggressively agitated with a mechanical stirrer, and the sample was placed in a hot bath to settle the sand. Using the wet screening method, the coarse sand was removed from a soil sample using a sieve of 200μm mesh. The prepared solution was then poured into the measuring cylinder with a stopper watch to record the time accurately. The solution was thoroughly mixed well before the pipette was used to the measure the 10 cm to determine the clay percentage (silt and clay percentage) of their sedimentation times.[18] The pipette method is used to quantify the proportion of silt and clay in a sample by measuring the weight of it. The textural triangle is a tool for determining the texture of soil.

2.7 Determination of ^{210}Po activity in soil

In a polythene bags, about 3 kg of individual samples were brought to the processing laboratory. For further investigation the soil samples were oven dried and moisture was removed for future analysis. The soil sample of approximately 20 g of each repeatedly leached (wet-ashing method) with 4N $HN0_3$ and filtered through WHATMAN 42. To oxidise organic materials in the samples, the acid is utilised. Organic matter was removed by gradually adding a 3:1 HNO_3 + $HClO_4$ solution until a white residue was formed. After that, the sample was placed in 1N HCl medium. A little amount of ascorbic acid was added to the solution to convert ferric ions to ferrous ions, preventing ferric ions from interfering with the deposition. A ^{209}Po tracer is used for the recovery. The ^{210}Po in the sample was assisted electrodeposited on a silver disc that was counted in the background. ZnS(Ag) Alpha scintillation counter coupled with photo multiplier tube for ^{210}Po with 30 percent of efficiency was used to measure the total count on both sides of the disk to find the ^{210}po activity.[19,20.21.22] The ^{210}Po activity concentration can be calculated using the following equation.

$$A = (S \pm SD) \times \frac{1000}{W} \times \frac{100}{E_p} \times \frac{100}{E}\ Bq\ kg^{-1} \qquad (3)$$

where, S represents the net count per second, SD represent the standard deviation, W represent the mass of the sample, E_p represent the plating efficiency taken as 90%, and E represent the efficiency of counter 30%.

2.8 Determination of ^{210}Pb activity in soil

The residual 0.5 M HCl was retained for one year after the first deposition of ^{210}Po to allow ^{210}Po in-growth from the ^{210}Pb in the solution to look for supported ^{210}Pb in the samples. The ^{210}Po activities were evaluated when the samples were re-plated. The ^{210}Pb activity was then determined by re-plating the solutions using silver discs and measuring the ingrown ^{210}Po. Two aliquots of the plating solutions for each sample were taken before the first and second platings for the measurement of total ^{210}Pb concentration. The activities of ^{210}Po and ^{210}Pb at the time of collection were determined by a series of corrections, including nuclide decay, ingrowth, chemical recoveries, detector backgrounds, and blank contamination following the methods[22]. Immediately after the chemical separation of the previously existing ^{210}Bi in the sample, the ingrowth of ^{210}Bi from the ^{210}Pb is commenced. The activity measurements of ^{210}Bi subsequent the chemical separation of a ^{210}Pb standard solution.[23] To calculate ^{210}Pb activity from recorded ^{210}Po activity, the Bateman equations were employed.[24] The second deposition revealed the samples ^{210}Pb level, and hence the degree to which the initial ^{210}Po was supported by its grandparent.

3 RESULTS AND DISCUSSION

Table 1 shows the activity of ^{210}Po and ^{210}Pb, and ratio of activity concentrations, in undisturbed sandy clay soil in the Belagavi district. The ^{210}Po activity concentrations ranged from 9.45 to 31.75 Bq kg^{-1}, with an average of 16.76 Bq kg^{-1}. The activity of ^{210}Pb ranged from 17.58 to 78.52 Bq kg^{-1}, with an average of 37.89 Bq kg^{-1}. The radionuclides activity ratio ranged from 0.39 to 0.54, with an average value of 0.4. The results show, that ^{210}Po and ^{210}Pb radionuclides concentration are not in equilibrium and the deposition of ^{210}Pb is much more compared to ^{210}Po in the soil. The disequilibrium between ^{210}Po and ^{210}Pb and the higher activity of ^{210}Pb indicates the presence of unsupported ^{210}Pb in the soil. The details of the ingrowth and decay calculation are dependent on the procedure followed. One common approach is that when^{210}Po is chemically separated, the solution containing the ^{210}Pb is retained and kept for an ingrowth period (usually several months). A subsequent determination of ^{210}Po from this solution is used to calculate the activity concentration of ^{210}Pb in the sample.[25,26,27] When analyzing for ^{210}Po it is important that the dry sample should not be heated above 150 °C to avoid loss of polonium and low recovery. Our emphasis is on the concentrations and distributions of ^{210}Pb and ^{210}Po in solid media, with focus on mineralized rocks and ores, and in anthropogenic materials resulting from the exploitation of natural resources. An assessment of the likely mineral hosts for both ^{210}Pb and ^{210}Po in critical geological environments, including ores and in non-nuclear industrial sources (technologically enhanced naturally occurring radionuclide material (TENORM). The clay and organic matter plays a vital role in fixing ^{210}Po and ^{210}Pb in soil in study area. The ^{210}Po activity distribution in fractionated dust and in the unattached phase might change dramatically depending on the meteorological conditions and local particle concentration.

Table 1 *^{210}Po and ^{210}Pb activity of soil in Belagavi district.*

Sl. No	**Activity ^{210}Po (Bq kg^{-1})**	**Activity ^{210}Pb (Bq kg^{-1})**	**Activity $^{210}Po/^{210}Pb$**
B1	25.65 ± 3.4	62.48 ± 14.8	0.41
B2	16.74 ± 1.6	38.24 ± 12.5	0.44
B3	12.48 ± 1.2	27.18 ± 5.2	0.46
B4	17.16 ± 1.3	34.26 ± 10.6	0.50
B5	18.25 ± 1.8	42.27 ± 14.3	0.43

B6	13.78 ± 0.9	30.60 ± 9.4	0.45
B7	13.67 ± 1.0	28.49 ± 6.7	0.48
B8	17.54 ± 2.4	32.56 ± 9.1	0.54
B9	16.45 ± 1.5	31.22 ± 10.0	0.53
B10	11.67 ± 1.1	24.28 ± 6.4	0.48
B11	19.57 ± 2.3	44.85 ± 14.7	0.44
B12	24.78 ± 2.7	58.34 ± 18.2	0.42
B13	10.46 ± 0.8	26.50 ± 5.8	0.39
B14	13.63 ± 1.5	28.65 ± 4.2	0.48
B15	12.82 ± 1	25.37 ± 4.7	0.51
B16	10.54 ± 0.8	22.74 ± 3.8	0.46
B17	12.45 ± 1.5	25.18 ± 3.2	0.49
B18	12.67 ± 1.8	26.22 ± 4.5	0.48
B19	14.52 ± 2.4	29.48 ± 3.4	0.49
B20	10.56 ± 1.3	21.33 ± 4.4	0.50
B21	13.76 ± 1.5	31.24 ± 5.2	0.44
B22	28.1 ± 3.2	68.35 ± 21.4	0.41
B23	22.45 ± 2.4	56.89 ± 17.6	0.39
B24	17.56 ± 1.8	42.58 ± 10.5	0.41
B25	26.23 ± 2.8	64.18 ± 22.13	0.41
B26	15.86 ± 1.9	40.57 ± 12.8	0.39
B27	31.75 ± 15.8	78.52 ± 28.6	0.40
B28	21.46 ± 4.8	54.12 ± 16.4	0.40
B29	9.45 ± 3.4	17.58 ± 1.6	0.54
B30	10.8 ± 2.8	22.46 ± 3.8	0.48
Minimum	**9.45**	**17.58**	**0.39**
Maximum	**31.75**	**78.52**	**0.54**
Mean	**16.76**	**37.89**	**0.46**

The physico-chemical parameters of soil in the Belagavi district are listed in Table 2. The electrical conductivity of soil in the study area ranges from 45 to 764 µSv cm. The percentage of organic matter in soil ranges from 3.70 to 15.57. The percentage of moisture in soil ranges from 1.84 to 15.72. Whereas the pH ranges from 4.12 to 7.52. Clay content is highest in the study area's maximal position, indicating the maturity of the profiles of soil. Clay plays an important role in the environment by acting as a natural scavenger of pollutants by taking up cation and anion either by ion exchange or adsorption or both.[28] The activity of these radionuclides is connected with the organic matter in order to comprehend the variance in the enrichment of both ^{210}Pb and ^{210}Po in surface soil (given in Table 2).

Table 2 *Physico-chemical parameters of soil in Belagavi district.*

Sl. No	Location	Moist %	OM%	pH	EC	Silt	Clay	Sand	Texture Nomenclat-ure
1	Telsang	12.83	5.56	6.58	210	0.20	26.37	73.43	Sandy Clay Loam
2	Athani	8.26	3.7	4.12	764	11.52	10.07	78.41	Sandy Loam
3	Ainapur	9.84	6.93	6.82	282	12.07	21.85	66.08	Clay loam
4	Chinchali	10.07	14.22	6.61	85	1.20	6.57	92.23	Sandy
5	Sadalaga	10.17	9.5	7.52	214	8.31	14.83	76.87	Sandy Loam
6	Raibag	12.36	8.36	7.21	258	38.72	18.12	43.16	Silty Loam
7	Harugeri	15.72	6.51	7.46	52	19.92	29.14	51.93	Clay loam
8	Nippani	7.16	7.94	6.92	85	18.49	15.44	64.06	Sandy Loam
9	Chikodi	14.45	4.52	6.16	549	32.17	32.67	35.16	Silty Clay Loam
10	Mudalagi	9.21	6.85	6.68	618	2.64	28.74	68.62	Sandy Clay Loam
11	Yadagudd	4.33	5.81	5.38	63	1.27	30.97	71.31	Sandy Clay Loam
12	Sankeshwar	8.28	15.57	4.94	580	11.03	21.47	67.5	Clay Loam
13	Hukkeri	4.35	8.31	4.56	145	3.2	9.23	87.57	Sandy Loam
14	Gogak	5.30	9.56	5.82	190	8.05	19.57	72.23	Sandy Loam
15	Bhutramhati	7.14	11.52	5.26	217	15.57	40.32	44.11	Clay
16	Ankalgi	6.02	13.22	6.74	107	42.46	11.62	45.91	Loam

17	Hirebudanur	6.87	10.07	7.2	45	0.28	3.32	95.46	Sandy
18	Katkol	11.59	7.85	7.17	84	8.31	14.83	76.87	Sandy Loam
19	Ramdurg	1.84	10.83	6.56	92	15.51	4.95	79.53	Clay Loam
20	Hoolikatti	9.32	8.26	5.64	325	40.79	3.35	55.86	Sandy Loam
21	Soudatti	10.43	9.84	6.52	50	18.13	16.89	64.98	Clay Loam
22	Hosur	6.68	12.43	7.51	128	35.45	24.59	39.96	Silty Loam
23	Bailhongal	14.22	10.17	5.82	113	18.47	14.85	66.68	Loam
24	Tigadi dam	9.50	12.36	5.26	138	19.38	39.24	41.38	Clay
25	Kittur	8.36	11.72	6.74	164	26.59	5.8	67.6	Silty Clay
26	Khanapur	6.51	7.16	5.74	124	7.85	13.22	79.02	Sandy Loam
27	Chorla	7.94	14.45	6.56	496	13.52	14.57	71.90	Loam
28	Kankumbi	4.52	9.21	7.43	520	20.54	31.72	47.73	Clay Loam
29	Kamsalli	4.61	7.60	7.10	86	20.46	23.62	55.89	Clay Loam
30	Sambra	4.78	10.80	5.40	432	0.20	26.37	73.43	Sandy Clay Loam

The ^{210}Po activity and soil organic matter content were associated with a good correlation coefficient of R = 0.85 (Figure 2). The ^{210}Pb activity and organic matter content having a significant correlation with R = 0.68 (Figure 3). The ^{210}Pb and ^{210}Po have a good association, with an R = 0.98 correlation coefficient (Figure 4). The ^{210}Po and ^{210}Pb activity varies from sampling locations to another location due to variation in the geological properties in soil, physico chemical parameters such as organic matter, pH, granular matrices, and immobility of soil.

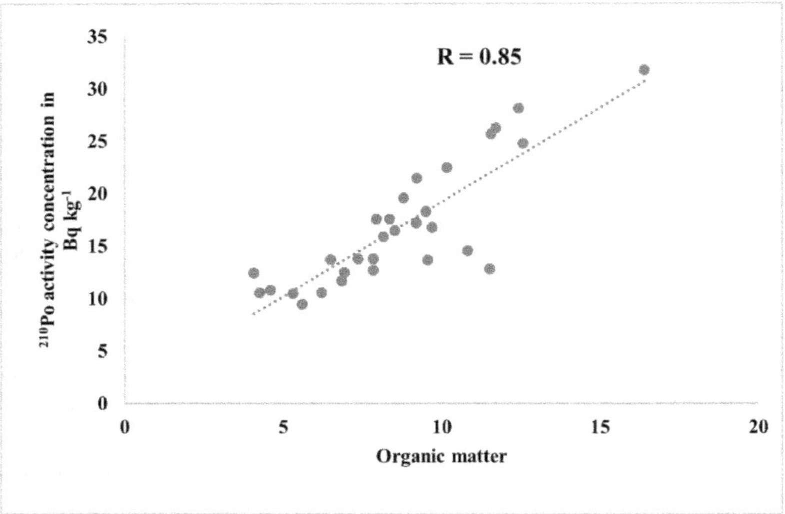

Figure 2 *Correlation of ^{210}Po with organic matter percentage.*

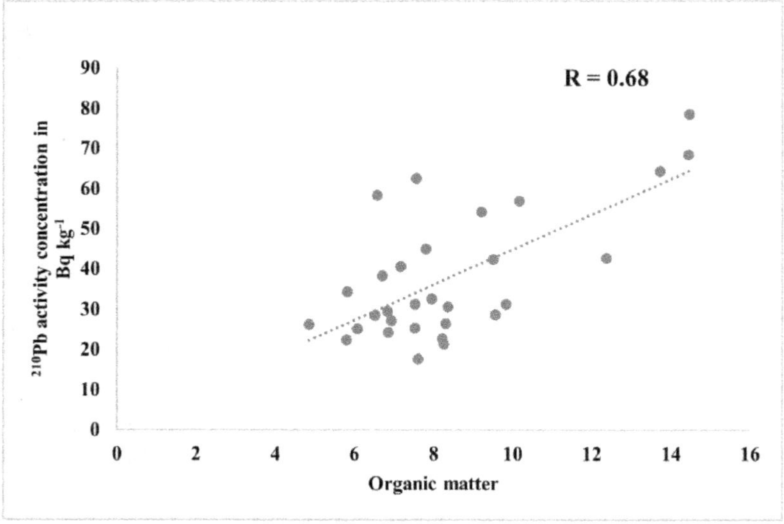

Figure 3 *Correlation of ^{210}Pb with organic matter percentage.*

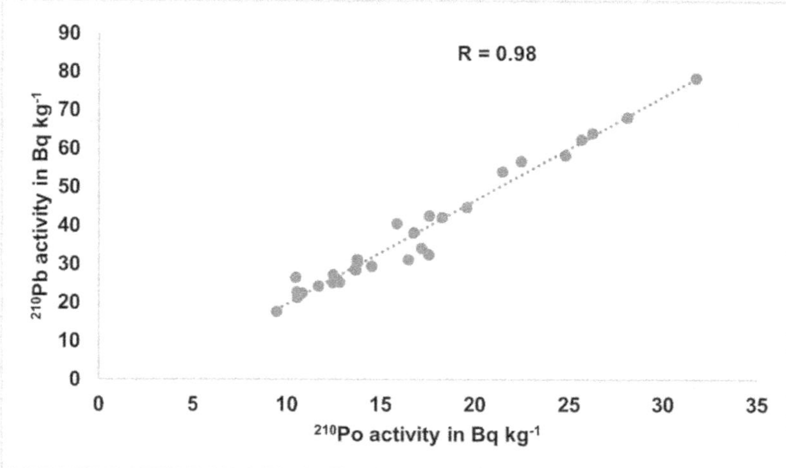

Figure 4 *Correlation of* ^{210}*Pb and* ^{210}*Po (Bq kg*$^{-1}$*) in soil.*

The activity of ^{210}Po and ^{210}Pb depends on soil and varies with soil type have reported that organic soils contain three times as much ^{210}Po as mineral soils. The variation of activity of ^{210}Po is found in following types of soils (black, brown and red). The polonium content in soil is very immobile and likely irreversible absorbed on organic colloids and clay content in soil. The minerals in soil were affected by the acidity present as pH in soil. Organic matter has a great influence on dynamics of radionuclides in soil because of its large cation exchange capacity.[29,30]

4 CONCLUSION

This research provides the first detailed report on distribution of ^{210}Po and ^{210}Pb activity and physico chemical effects of undisturbed alluvial clay soil in Belagavi district, Karnataka, India. The average of ^{210}Po activity in soil samples ranged from 9.45 to 31.75 Bq kg^{-1}, with an average of 16.76 Bq kg^{-1}. Whereas, the ^{210}Pb activity range from 17.58 to 78.52 Bq kg^{-1}, with an average value of 37.89 Bq kg^{-1}. The higher activity of ^{210}Pb in the soil when compared to ^{210}Po can be attributed to the difference in the geochemical pathways of these two radionuclides. Organic matter plays an important role in the accumulation and enrichment of ^{210}Po and ^{210}Pb in soil. The limited availability of ^{210}Po and ^{210}Pb radionuclides in studied soil samples led to the conclusion that their presence in the environment makes ingesting and inhalation more harmful. Therefore, ^{210}Po activity in both phases and other alpha-emitting radionuclides should be monitored more preciously. of ^{210}Po, which is released to the atmosphere mostly in winter season. In such areas, ^{210}Po activity should be carefully monitored in the air, as well as in all types of environmental samples, including soil, water, and locally produced food.

5 ACKNOWLEDGMENTS

The authors acknowledge the support from DST, Govt of India under the FIST program, to augment of research facilities in the department. One of the authors (KVD) is grateful to the SC/ST Cell Mangalore University for the fellowship during the research studies.

6 DISCLOSURE STATEMENT

No potential conflict of interest was reported by any of the authors.

7 FUNDING

This research did not receive any specific grant from funding agencies in the public, commercial, or not-for-profit sectors.

8 REFERENCES

1. United Nations Scientific Committee on the Effects of Atomic Radiation, Vol. I: Sources. United Nations Publications, New York, 2000.

2. NCRP, National Council on Radiation Protection and Measurement No.45, *Soil Radioactivity*, 1975.

3. E. Ansoborlo, P. Berard, C. Den Auwer, R. Leggett, F. Menetrier, A. Younes, G. Montavon and P. Moisy, *Chem. Res. Toxicol,* 2012, **25**,1551.

4. K. Matthews, C. Kim and P. Martin, *Appl. Radiat. Isot,* **65**, 267.

5. M. Edwards, S. Tryantafyllidou and D. Best, *Environ. Sci. Technol,* 2009, **43**, 1618.

6. United Nations. Scientific Committee on the Effects of Atomic Radiation. *UNSCEAR Report to the General Assembly, with Scientific Annexes (Vol. 2).* United Nations Publications, 2011.

7. E. Sert, A. Uğur, B. Özden, M. Murat Saç and B. Camgöz B, *J Environ. Radioact.* 2011, **102**, 535.

8. O. Kılıc¸ M. Belivermis and Y, Cotuk, *Mar. Pollut. Bull*, 2014, **80**, 325.

9. V. Kamalakar, P. Vinutha, C. Kaliprasad and Y. Narayana, *Environmental Forensics*, 2022, DOI: 10.1080/15275922.2022.2125113

10. V. Kamalakar, P. Vinutha, C. Kaliprasad and Y. Narayana, *International Journal of Environmental Analytical Chemistry*, 2022, DOI: 10.1080/03067319.2022.2027932

11. K. Prabhavati, G. Dasog, K. Sahrawat, P. Patil and S. Wani, *Journal of the Indian Society of Soil Science,* 2017, **65(1)**, 1.

12. IAEA, Guidebook, A. Measurement of Radionuclides in Food and the Environment. Vienna: *International Atomic Energy Agency.* Retrieved from https://www. iaea.org/publications/1398/measurement-of-radionuclides-in-food-and-the-environment. 1989.

13. N. Chieco, *Volume 2, 28 (No. HASL-300-Vol. 2-28. ed.). USDOE Environmental Measurements Lab., New York, NY (United States).* http://www.iaea.org/inis/NCL%20/_Public/30/000/30000468, 1997.

14. R. Miller and D. Kissel, *Soil Science Society of America Journal,* 2010, **4(1)**, 310.

15. M. Prakash, C., Kaliprasad and Y. Narayana, *Journal of Radioanalytical and Nuclear Chemistry,* 2018, **315(2)**, 315.

16. M. Prakash, C., Kaliprasad and Y. Narayana, *Journal of Radiation Research and Applied Sciences,* 2010, **10(1),** 57.

17. Y. Narayana, P. Shetty and K. Siddappa, *Appl. Radiat. Isot.* 2009, **64,** 396.

18. K. Rajashekara, V. Prakash and Y. Narayana, *Journal of Radioanalytical and Nuclear Chemistry,* 2019, **322(1),** 199.,

19. P. Vesterbacka, *Boreal Environ. Res.* 2007, **12(1),** 11.

20. N. Sethy, A. Sutar, P. Rath, V. Jha, P. Ravi and R. Tripathi, *Journal of Radiation Research and Applied Sciences,* 2015, **8(4)**, 590.

21. N. Venunathan and Y. Narayana, *Journal of Radiation Research and Applied Sciences,* 2016, **9(4)**, 392.

22. S. Rigaud, G. Stewart, M. Baskaran, D. Marsan and T. Church, US GEOTRACE GA03 section, Deep-Sea Res. Pt. II, 116, 60, 2015

23. N. Carfi and R. Lonati, *In Proceedings of the First International Congress of Radiation Protection*, 1097, Pergamon, 1968.

24. V. Prakash, V. Vineethkumar, K. Rajashekhara and Y. Narayana, *Radiation Protection and Environment,* 2019, **42(3)**, 102.

25. G. Jia, M. Belli, M. Blasi, A. Marchetti, S. Rosamilia and U. Sansone, *J. Radioanal. Nucl. Chem.* 2001, **247,** 491.

26. P. Martin and G. Hancock, *Supervising Scientist Report 180,* AGPS, Canberra, 2004.

27. L. Johansson, PhD Thesis, Gottfried Wilhelm Leibniz Universität Hannover, 2004.

28. J. Cetnar, *Annals of Nuclear Energy,* 2006, **(7)**, 640.

29. S. Gupta and K. Bhattacharyya, *Journal of environmental management,* 2008, **87(1)**, 46.

30. M. Aslani, S. Akyil, S. Aytas, G. Gurboga and M. Eral, *Radiation measurements,* 2005, **39(2)**, 129.

CHAPTER 13 ANALYSIS OF THE SORPTION OF Tc(IV) TO SOME CLAY MINERALS WITH REFERENCE TO RADIOACTIVE WASTE DISPOSAL

N. D. M. Evans[a*] and R. J. Hallam[b]

[a]School of Science and Technology, Nottingham Trent University, Clifton Campus, Nottingham, NG11 8NS, UK, [b]Loughborough University, Epinal Way, Loughborough, UK

*nick.evans@ntu.ac.uk

1 INTRODUCTION

The United Kingdom has recently formed a new agency to manage the country's radioactive wastes. Nuclear Waste Services (NWS) have merged the organisations, the Low-Level Waste Repository, Radioactive Waste Management (RWM), and the Nuclear Decommissioning Authority group's Integrated Waste Management Programme together under one umbrella. This has created an organisation focused on the management of the UK's nuclear waste, safely and securely for generations to come.

NWS will manage the delivery of the UK's first Geological Disposal Facility (GDF). This involves working with communities to find a suitable site and a willing community to host a GDF.[1] In order for this to be successful, the Performance Assessment (PA) and Environmental Safety Cases (ESC)will need to be of sufficient calibre to meet the requirements of Government, the regulators, such as the Environment Agency, and the local population near any possible site.

This requires robust experimental data which will inform the PA of the planned GDF. One of the key radionuclides for PA is ^{99}Tc. ^{99}Tc is a radioactive isotope which is formed in nuclear reactors as a by-product (~6%) of the fission of ^{235}U. It has a half-life of 2.13 x 10^5 years and will ultimately have to be disposed of in a GDF. Within the highly alkaline conditions of the GDF, and at circumneutral pH values, the aqueous chemistry of reduced technetium is dominated by $TcO(OH)_2(aq)$.[2] This species has been shown to be capable of surface complexation reactions at either one or both of the hydroxyl groups.[3]

The purpose of this research was to ascertain the chemistry of low-oxidation state technetium within a radioactive waste repository context. A major area of interest is the sorption of Tc(IV) to the surface of various clay minerals likely to be in the vicinity of, or inside, the near- and far-fields of the proposed GDF. Sorption experiments were conducted at pH values of 13.3, which is the initial pH expected in a GDF and pH 12.5 which is the expected equilibrium pH of the GDF after many years. The Alkaline Disturbed Zone (ADZ), the geological area in the immediate vicinity of the GDF, was also investigated by conducting experiments at pH 10 and the final pH value investigated was pH 7, representing the geosphere at a greater distance away from the GDF and the ADZ, and also at pH 3 for those scenarios where acidic wastes are considered. The clay minerals examined were, bentonite, illite, montmorillonite and a smectite-illite mixed layer.

The JCHESS geochemical speciation code was used to model sorption behaviour on the basis of monodentate and/or bidentate binding as a result of one or both of the hydroxyl groups being involved in the sorption reactions.[4]

2 EXPERIMENTAL

2 mL of a solution of ammonium pertechnetate (NH_4TcO_4) was poured into a three-necked round-bottomed flask and made up to 200 mL with 18 MΩ de-ionised water producing a pertechnetate (TcO_4^-) concentration of 6.79 x 10^{-7} mol L^{-1}. Sodium hydroxide (NaOH) was then added to produce solutions of the required pH, with a target of producing a range of final solutions from approximately pH 3 to 13.3. The solutions were then purged for 2 hours with oxygen-free (< 1 ppm) nitrogen to remove any dissolved oxygen and/or carbon dioxide. A vitreous carbon cathode and a platinum wire anode were then placed into the cell in a nitrogen glove box with O_2 content less than 1 ppm. A six volt electro-potential was used to reduce the pertechnetate to Tc(IV). Complete reduction from Tc(VII) to Tc(IV) was completed in a few hours. This allowed the production of Tc(IV) aqueous solutions of a concentration of approximately 1 x 10^{-9} mol L^{-1}. Solutions with Tc(IV) concentrations of 1 x 10^{-10} and 1 x 10^{-11} mol L^{-1} were also prepared, to ensure that precipitation of TcO_2(am) did not occur in any of the experiments. At such concentrations, the specific activity of ^{99}Tc is not high enough to be measured by liquid scintillation counting, so the solutions were spiked with a few kBq of ^{95m}Tc prepared by the University of Wisconsin. ^{95m}Tc is a gamma photon emitter with a half-life of 61.96 days,[5] and with measurable gamma photon energies of 204.1 and 582.1 keV.[6]

Ground clay minerals (0.1 g, <180 μm) were weighed into 20 cm^3 polypropylene vials. Tc(IV) solutions were added and left for at least one week to equilibrate in a nitrogen glove box, on a flatbed stirrer. This method was performed, in at least triplicate, for each clay mineral using each of the different pH solutions. The pH was measured at the end of the equilibration phase, to allow for the buffering effects of the clay minerals. After equilibration, a 5 mL aliquot of the supernatant solution was removed from the vials through a 0.22 μm syringe filter with approximately 2 mL being weighed into a gamma vial and counted using a Packard Cobra II Autogamma counter. The remaining solution was discarded to waste. No measurable wall sorption, or sorption to the filters was detected. The samples were left to equilibrate once the solution had been added as it is evident that there is a natural buffering capacity of the solid which alters the pH from that of the initial solution of pertechnetate.

To determine the surface area for each mineral, approximately one gram of each clay was dried at 60 C under vacuum for several hours. The clay was weighed accurately to 4 decimal places. The BET surface area was measured on a Micromeritics Gemini VI physisorption system following the method of Kemp *et al.* at the British Geological Survey.[7] Gemini software was used to obtain the BET surface areas. The cation exchange capacities of each clay were determined by titration using a Metrohm Titrando 857 equipped with a Metrohm Dosino 800 dosing unit. Around 0.3 g of solid was used with 60 mL of de-ionised water (18 MΩ) to maintain the solid:solution ratio at the same level as in the sorption experiments and ensure that the solution level sufficiently covered the pH probe. The solution was stirred and 2 mol L^{-1} HCl added drop-wise until the pH stabilised at approximately 2 - 3 (fully protonated). At this point, 0.1 mol L^{-1} sodium hydroxide solution was added at a rate of 0.1 mL min^{-1} until the pH had risen and stabilised above 11 - 12. The points of zero charge for each clay mineral were determined to be: montmorillonite pH = 3.4 ± 0.2; bentonite pH = 1.8 ± 0.3; smectite pH = 3.4 ± 0.3; illite-smectite mixed layer = 3.5 ± 0.2.

3 RESULTS AND DISCUSSION

The aqueous chemistry of technetium in its +4 oxidation state is dominated, in all but highly acidic solutions (< pH 2), by the species $TcO(OH)_2(aq)$, as can be seen in Figure 1.[2]

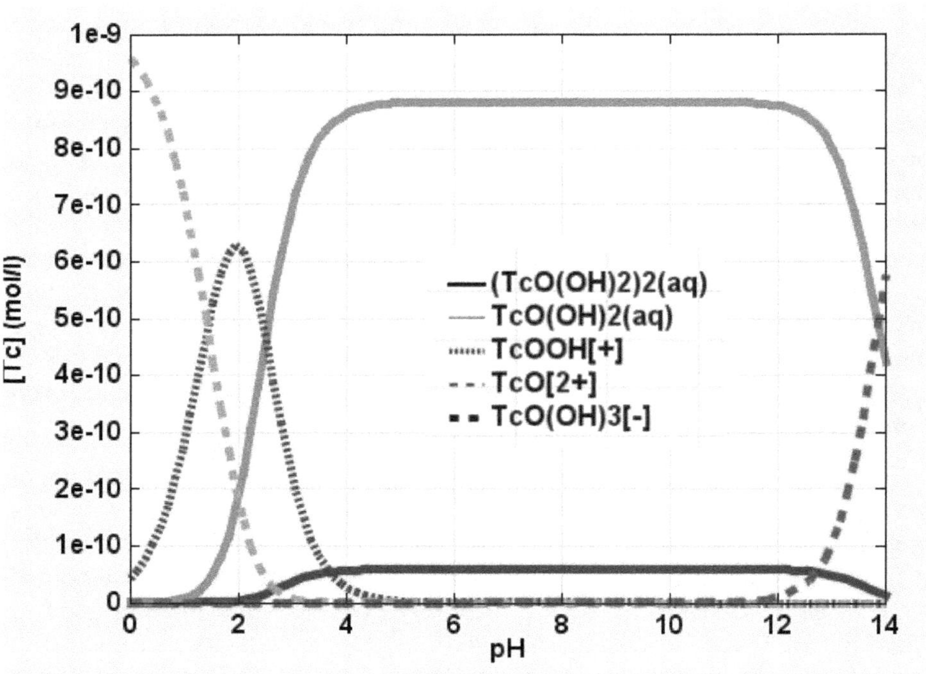

Figure 1 *Speciation of Tc(IV) as a function of pH, from the JCHESS code. (Total Tc(IV) concentration 10^{-9} mol L^{-1}).*

This species has been shown to be capable of surface complexation reactions by exchange of one or two of the hydroxyl groups[3]. Therefore, the sorption behaviour of Tc(IV) observed in this study has been modelled on the basis of mono- or bidentate surface complexes, following Grambow *et al.*[3] The geochemical speciation code JCHESS[4] was used for performing the surface complexation modelling.

3.1 Montmorillonite

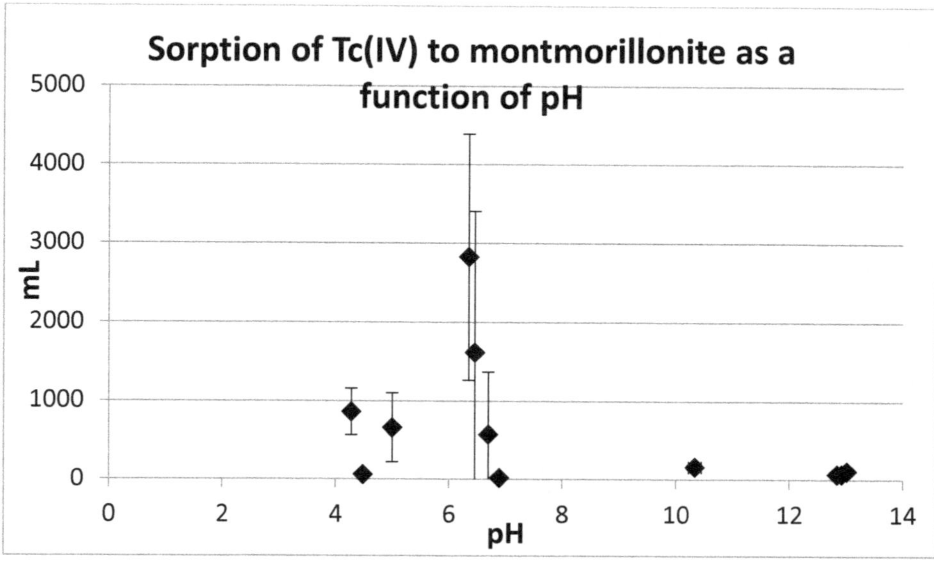

Figure 2 *The sorption of Tc(IV) as a function of pH for montmorillonite at ambient temperature in an oxygen free atmosphere (<1 ppm). Initial Tc(IV) concentrations 10^{-9} to 10^{-11} mol L^{-1}. Solid to liquid ratio 1:200. Data uncertainties are shown as plus/minus one standard deviation.*

The peak in the distribution coefficient (R_d) values was observed to be at pH = 6.4. This set of sorption experiments follow a pattern whereby the distribution coefficient rises with pH until near neutral conditions before falling rapidly once the pH tends towards alkalinity. This can be seen in detail in Table 1 below.

Table 1 *Partition coefficients, R_d, (mL g^{-1}) for the sorption of Tc(IV) to montmorillonite at ambient temperature in an oxygen free atmosphere (<1 ppm). Initial aqueous Tc(IV) concentrations 10^{-9} to 10^{-11} mol L^{-1}. Solid to liquid ratio (1:200). Data uncertainties are shown as ± one standard deviation.*

pH	R_d (mL g^{-1})	pH	R_d (mL g^{-1})
4.3 ± 0.03	860 ± 290	6.9 ± 0.02	19 ± 2
4.5 ± 0.05	60 ± 3	10.3 ± 0.11	162 ± 1
5.00 ± 0.05	660 ± 440	12.8 ± 0.03	72 ± 13
6.4 ± 0.03	2830 ± 1560	12.9 ± 0.02	70 ± 11
6.5 ± 0.02	1620 ± 1790	13.0 ± 0.01	110 ± 7
6.7 ± 0.02	580 ± 800		

It has proved possible to fit the data for montmorillonite, by using both the mono- and bi-dentate complexation equations. This model has, however, produced a much sharper peak than previous models of this type that have been used, such as in Hallam and Evans (2011). [8,9] This plot can be seen below

(Figure 3) along with the equations and parameters used. As can be seen, the model fits well with the experimental data obtained from the laboratory experiments. (See Table 2).

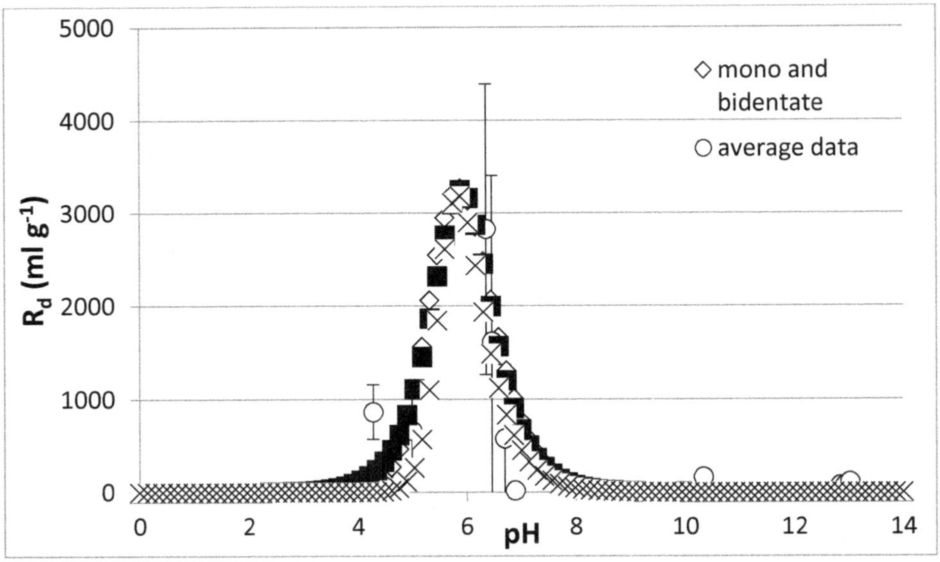

Figure 3 *Modelling the sorption of Tc(IV) as a function of pH for montmorillonite.*

Table 2 *Surface Complexation Reactions and constants used in modelling the sorption of Tc(IV) to montmorillonite at varying pH. The designation Mont-OH indicates the surface groups controlling the sorption to montmorillonite.*

Surface Complexation Reaction used in modelling	**Log K**
Mont-OH + H^+ $\leftrightarrow$ Mont-OH_2^+	4.5
Mont-OH $\leftrightarrow$ Mont-O^- + H^+	-5.8
2Mont-OH + TcO^{2+} + $2H_2O$ $\leftrightarrow$ $(Mont\text{-}O)_2TcO(OH)$ + $3H^+$	-3.4
Mont-OH + TcO^{2+} + H_2O $\leftrightarrow$ Mont-OTcO$(OH)^-$ + $2H^+$	-5.2

3.2 Illite

See Figure 4 for the sorption of Tc(IV) as a function of pH.

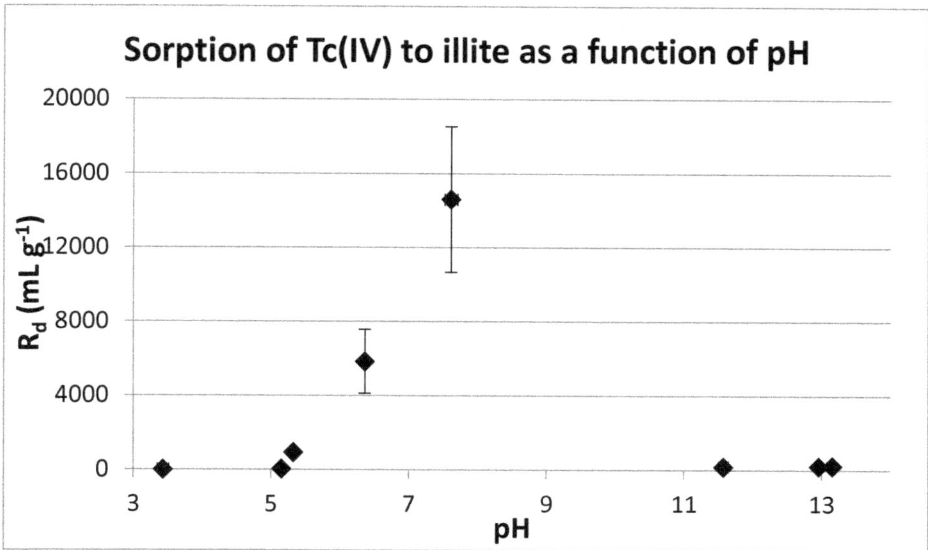

Figure 4 *Tc(IV) sorption to illite as a function of pH at ambient temperature in an oxygen free atmosphere (<1 ppm). Initial Tc(IV) concentrations 10^{-9} to 10^{-11} mol L^{-1}. Solid to liquid ratio (1:200). Data uncertainties are shown as ± 1 standard deviation.*

Measured values for the distribution coefficient for the sorption of Tc(IV) to illite are shown in Table 3 below. The peak measured R_d value of 14 600 mL g^{-1} at near neutral pH follows the general trend to that of the previous solids tested up to this point despite the solids investigated being very different, i.e. igneous rocks and cements, as well as the montmorillonite data above.[8,9]

Table 3 *Partition coefficients, R_d, (mL g^{-1}) for the sorption of Tc(IV) to illite at ambient temperature in an O_2 free atmosphere (<1 ppm). Initial aqueous Tc(IV) concentrations were 10^{-9} to 10^{-11} mol L^{-1}. Solid to liquid ratio (1:200). Data uncertainties are shown as ± 1 S.D.*

pH	R_d (mL g^{-1})	pH	R_d (mL g^{-1})
3.4 ± 0.09	8.2 ± 3.8	7.6 ± 0.10	14600 ± 390
5.2 ± 0.04	29.6 ± 5	11.6 ± 0.07	170 ± 98
5.3 ± 0.03	910 ± 70	13.0 ± 0.00	214 ± 6
6.4 ± 0.04	5840 ± 1730	13.2 ± 0.01	222 ± 85

The sorption of Tc(IV) to illite has proven difficult to model to the same level of accuracy as some of the other solids that have been experimentally tested and modelled in these studies. The data again show that there is a trend – R_d rises as pH rises and then decreases once alkaline conditions become dominant. The lower and upper sorption edges can be found at pH ~5 and ~10 respectively. As can be observed in Figure 5, the closest fit to the lower sorption edge has been found to be the initial slope of the monodentate binding model.

The lower sorption edge can be modelled using both mono- and bi-dentate modelling equations, but in both cases the upper sorption edge does not match that of the experimental data. However, when the model is used to fit the upper edge the fit then moves away from the lower sorption edge. When both models are combined it has found that the mono-dentate model dominates the shape of the model and once the effect of the bi-dentate has dropped below that of the mono-dentate model then the curve stays true to that of the mono-dentate model.

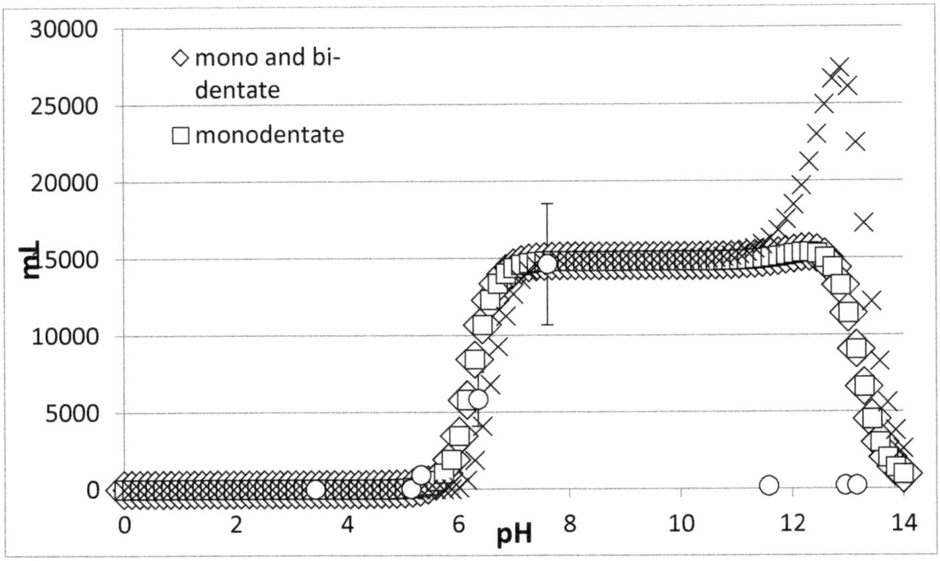

Figure 5 *Sorption of Tc(IV) as a function of pH for illite at ambient temperature in an oxygen free atmosphere (<1 ppm). Initial Tc(IV) concentrations 10^{-9} to 10^{-11} mol L^{-1}. Solid to liquid ratio 1:200. Data uncertainties are shown as ± 1 S.D.*

3.3 Smectite/illite mixed layer

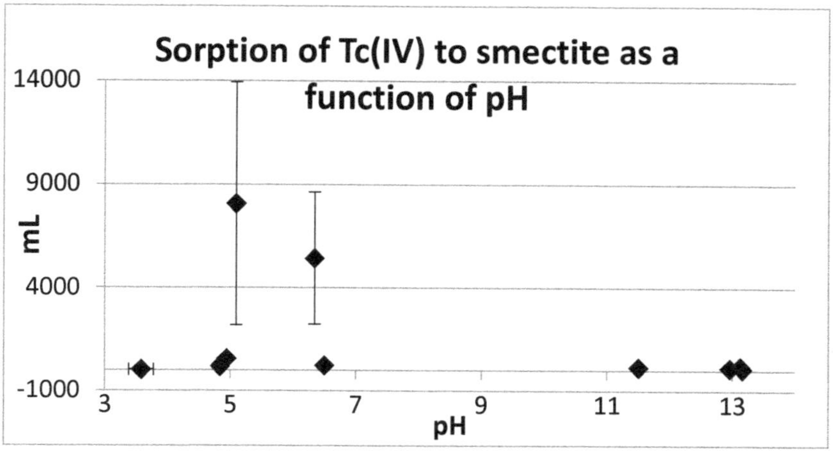

Figure 6 *Modelling the sorption of Tc(IV) to smectite as a function of pH.*

The degree of sorption of Tc(IV) to the smectite-illite mixed layer, was much lower than for that of illite alone (See Figure 6). The highest R_d value was found to be 8 070 mL g^{-1}, see Table 4, and was also found at a slightly lower pH, i.e. 5.1, compared to other solids that have been previously investigated where sorption generally peaks at a pH much closer to neutral conditions.

Table 4 *Distribution coefficients, R_d, (mL g^{-1}) for sorption of Tc(IV) to smectite-illite mixed layer at ambient temperature in an O_2 free atmosphere (<1 ppm). Initial aqueous Tc(IV) concentrations 10^{-9} to 10^{-11} mol L^{-1}. Solid to liquid ratio (1:200). Data uncertainties shown as ± 1 S.D.*

pH	R_d (mL g^{-1})	pH	R_d (mL g^{-1})
3.6 ± 0.20	28 ± 22	6.5 ± 0.00	238 ± 100
4.8 ± 0.02	187 ± 35	11.5 ± 0.03	150 ± 67
4.9 ± 0.01	560 ± 95	13.0 ± 0.00	124 ± 71
5.1 ± 0.05	8070 ± 5870	13.1 ± 0.00	210 ± 40
6.4 ± 0.04	5440 ± 3200	13.2 ± 0.00	74 ± 3

When modelling the experimental results for illite-smectite mixed layer the same shape curves as illite were found to fit the data, though the R_d values are considerably lower. This would suggest that although illite is the major contributor to the sorption of Tc(IV) the smectite component has significantly reduced the sorption of Tc(IV). Figure 7 shows that the monodentate section of the modelling curves fits the experimental data at the lower sorption edge very well and also at the extreme alkaline conditions. However, the data measured at pH ~11 could not be modelled so closely.

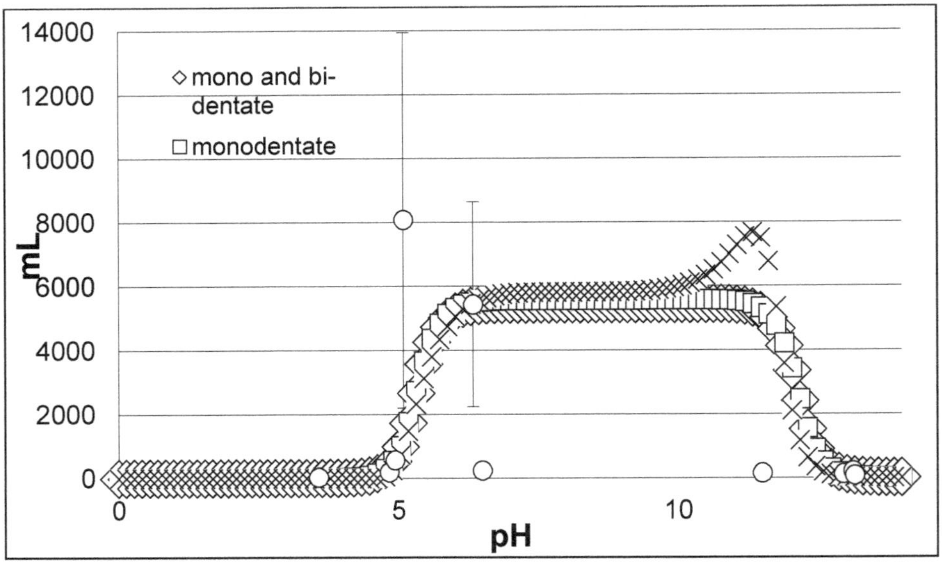

Figure 7 *The sorption of Tc(IV) as a function of pH for illite-smectite mixed layer at ambient temperature in an oxygen free atmosphere (<1 ppm). Initial aqueous Tc(IV) concentrations were 10^{-9} to 10^{-11} mol L^{-1}. Solid to liquid ratio 1:200. Data uncertainties are shown as ±1 S.D.*

3.4 Bentonite

Sorption of aqueous Tc(IV) to bentonite was highest at pH 8.3, at a concentration of 10^{-10} mol L^{-1} (See figure 8). This peak is at a considerably higher pH than for the other clay minerals. The R_d values for bentonite are similar to those for montmorillonite and smectite, but it proved impossible to model the data using the mono- and bidentate surface complexation model used above. This suggests that sorption to bentonite is more complicated that for the other clay minerals.

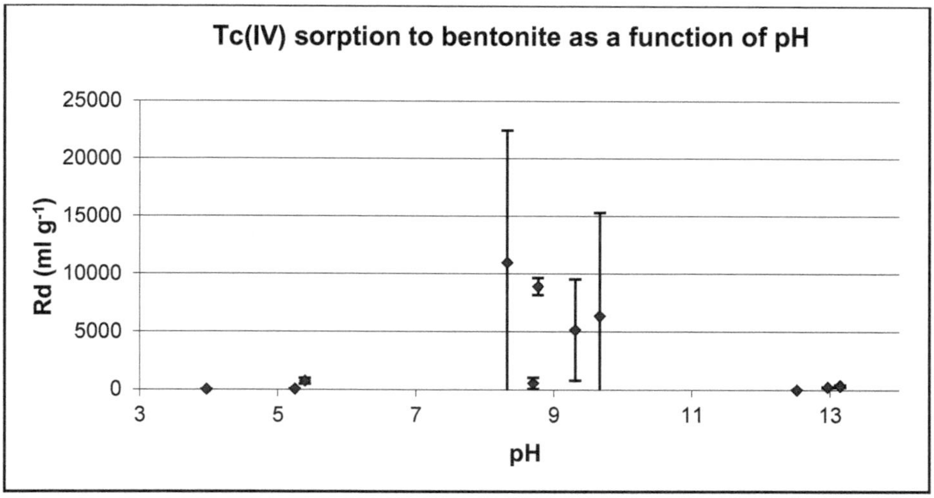

Figure 8 *The sorption of Tc(IV) as a function of pH for bentonite at ambient temperature in an oxygen free atmosphere (<1 ppm). Initial aqueous concentrations of Tc(IV) were 10^{-9} to 10^{-11} mol L^{-1}. Solid to liquid ratio 1:200. Data uncertainties are shown as ±1 S.D.*

3.5 Overview

The distribution coefficient (R_d) data for all four investigated clay minerals across the pH range is shown in Figure 9.

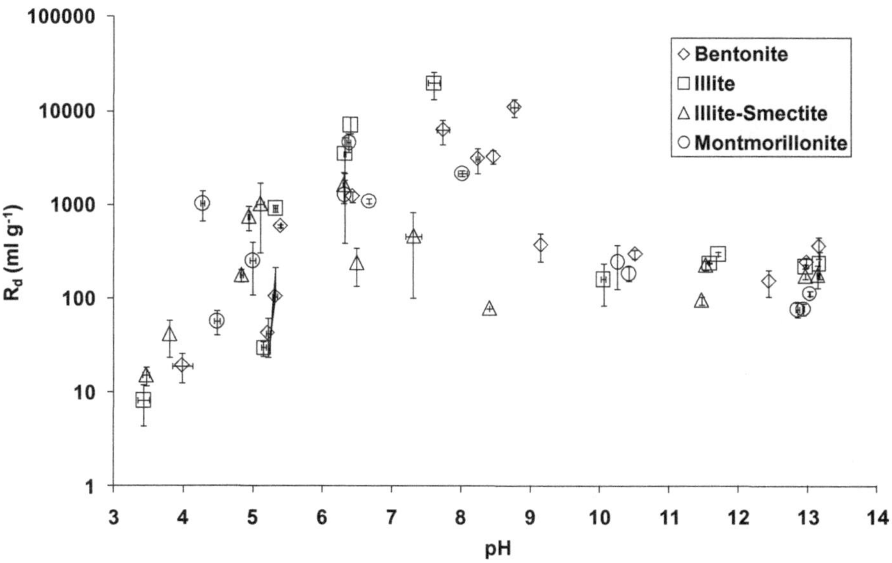

Figure 9 *R_d as a function of pH for Tc(IV) sorption to 4 sample clays.*

4 CONCLUSIONS

Comparing the sorption properties for Tc(IV)(aq) for the clay minerals shows that the measured distribution coefficient, R_d, for Tc(IV) to montmorillonite was much lower than those of the other clays that were used in this study. The R_d values for montmorillonite were observed to peak at 2 830 mL g^{-1} whereas those for illite observed to be much higher peaking at approximately 15 000 mL g^{-1}. The illite-smectite mixed layer peak R_d value was observed to be 8 070 mL g^{-1} and bentonite, slightly higher at ~12 000 mL g^{-1}.

In all the measured R_d values, which are heavily dependent on pH, were in a range from 70 to 16000 mL g^{-1}, with maximum sorption occurring in near-neutral pH range, with an increase sometimes occurring above pH 13. In terms of modelling, bidentate surface complexation constants in hyperalkaline region, and a combination of mono- and bidentate below *ca.* pH 11.5 gave best agreement. These data will inform the performance assessment for the near-field and far-fields of the UK's planned higher-activity wastes GDF.

5 ACKNOWLEDGEMENTS

The authors are grateful for the help of Jerry Nickles (University of Wisconsin) for preparation of ^{95m}Tc and the British Geological Survey (BGS) for assistance with the BET surface area measurements.

6 REFERENCES

1. https://www.gov.uk/guidance/gdf-geological-disposal-facility 26.11.2022.
2. P. Warwick, S. Aldridge, N. Evans and S. Vines, *Radiochimica Acta*, 2007, **95(12)**, 709.
3. B. Grambow, M. Fattahi, G. Montavon, C. Moisan and E. Giffaut, *Radiochimica Acta*, 2006, **94(9-11)**, 627.
4. J. Van der Lee, "JChess 2.0", Ecole des Mines de Paris, Centre d'Informatique Géologique, http://chess.ensmp.fr jchess (2000).
5. T. Szegedi, Á. Tóth, G. Kiss and Gy. Gyürky, *The European Physical Journal A*, 2020, **56(7)** 182.
6. www.LNHB.fr (Laboratoire National Henri Becquerel) accessed 25.11.2022.
7. S. Kemp, G. Turner, D. Wagner, *British Geological Survey Internal Report*, 2009, IR/08/086.
8. R. Hallam, N. Evans and S. Jain, *Mineralogical Magazine*, 2011, **75(4)**, 2439.
9. R. Hallam and N. Evans, *Proceedings of Waste Management Symposium* 2011, Published on CD – 11129.

CHAPTER 14 PRECISE MEASUREMENT OF TRACE ACTINIDES USING PHOENIX TIMS

Z. Palacz[a], S. Yardley[a], M. Hockley[a] and S. Guilfoyle[a*]

[a]Isotopx Ltd, Middlewich, Cheshire, UK

*Stephen.Guilfoyle@isotopx.com.

1 BACKGROUND

When measuring very small ion signals (<10,000 cps), in sub-picogram samples of actinides, ion counting detectors are necessary due to their low noise compared to that of a Faraday detector. Ion counters, however, still have a non-zero noise associated with them, known as the dark noise, which is typically several counts per minute (cpm).

Additionally, there are sources of measurement error present that are not of concern with Faraday detectors, such as deadtime/linearity, peak flatness, and gain stability. Isotopx developments in the MICX electronics aims to address these issues to produce higher quality data at lower count rates. See Figure 1.

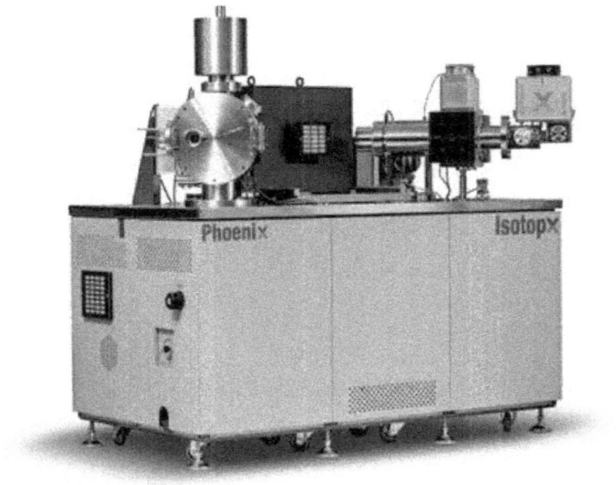

Figure 1 *The Isotopx Phoenix.*

2 ARE THE PEAKS FLAT?

Peak flatness is a source of error in ion counting measurement. A slope on the flat can result in an apparent gain change depending on where on the peak flat a measurement is made. The peak flat is shown in Figure 2 below, and scans across all ion counters are shown in Figure 3. The peaks are indeed flat.

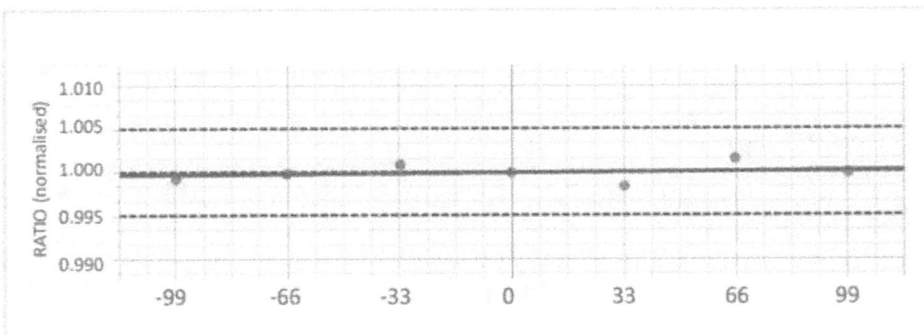

Figure 2 *Typical peak is flat to better than 0.1% over +/-100 ppm of mass.*

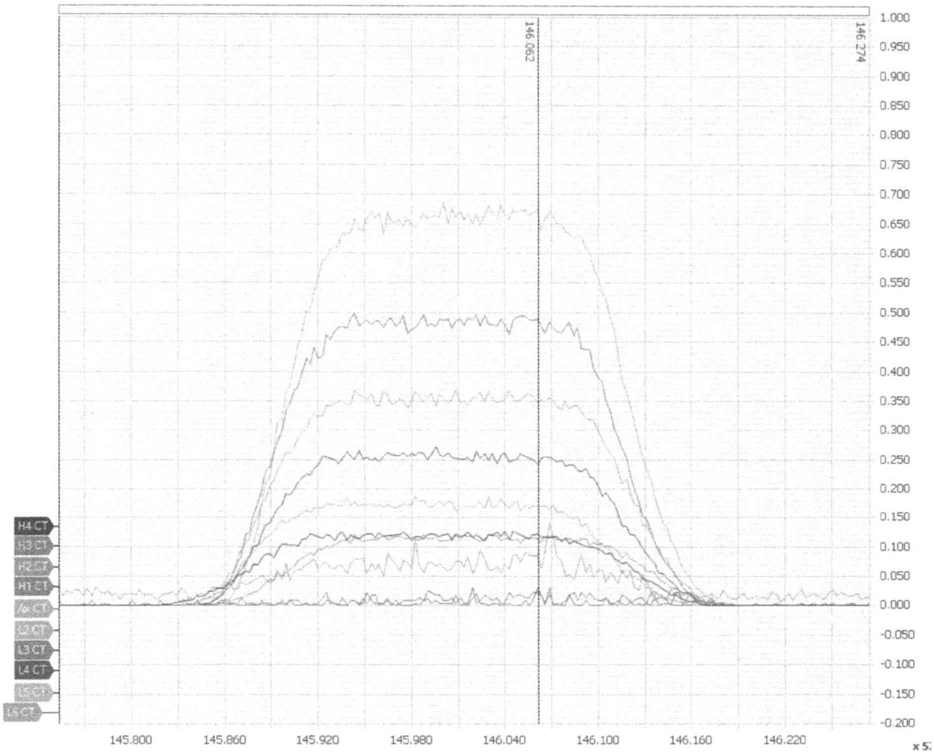

Figure 3 *Coincident beams on all 10 ion counters. This demonstrates coincidence across the instrument focal plane.*

3 WHAT ABOUT DARK NOISE?

Dark noise is a measure of the counts registered on an ion counter when no beam is present. Ten channels of dark noise data were collected over several hours, shown below (Table 1). In all cases the dark noise was less than 0.6 counts per minute (0.01 cps).

Table 1 *Ion counter dark noise [counts per minute]. The average of all channels is less than 0.5 cpm.*

L6	L5	L4	L3	L2	Ax	H1	H2	H3	H4
0.49	0.50	0.52	0.37	0.45	0.57	0.38	0.45	0.49	0.42

4 ARE THESE ION COUNTERS LINEAR?

Ion counting detectors are inherently non-linear. This can be corrected by applying a deadtime. The stability of this deadtime and the resultant correction is critical; nonlinearity will be observed as a variation of the measured isotope ratio with incident beam intensity.

The linearity of the ten ion counters was determined by analysing the $^{234}U/^{238}U$ ratio of a uranium standard (NBS U500). The analysis was performed with a ^{238}U intensity of 5 x 10^4 cps to 3 x 10^5 cps. Figure 4 shows typical data, demonstrating linearity to better than 0.1% 1RSD. The ion counters are linear.

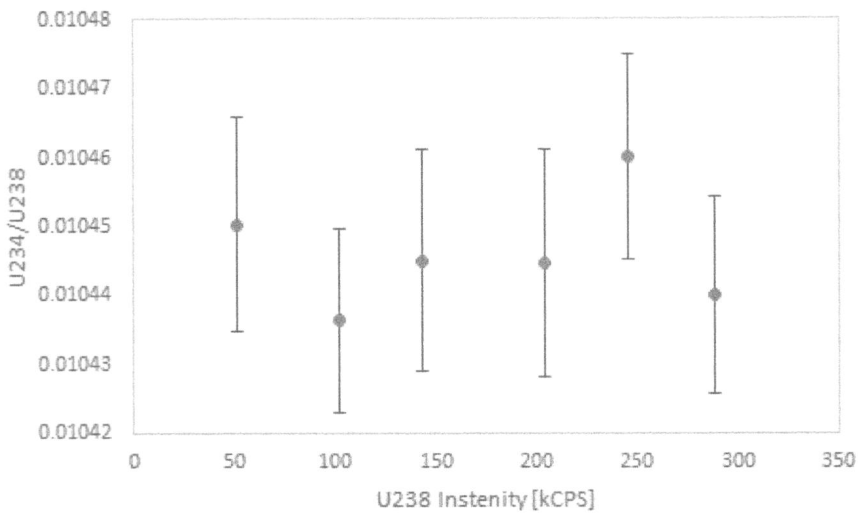

Figure 4 *Isotope ratio obtained as a function of incident beam intensity. Linearity is better than 0.1% 1RSD.*

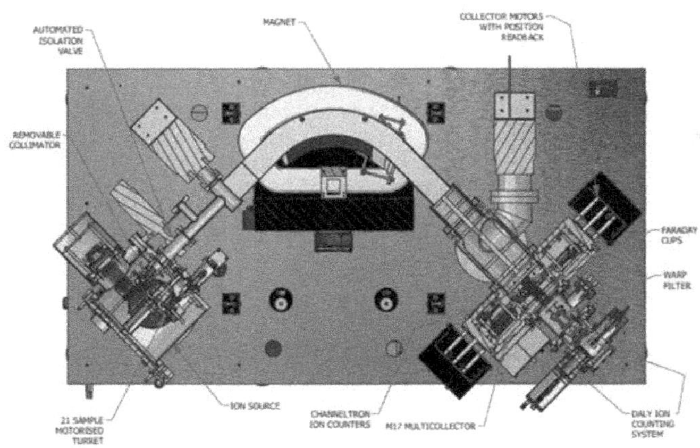

Figure 5 *Instrument schematic.*

5 ARE THE GAINS STABLE?

Gain stability of ion counting detectors is another factor. The $^{235}U/^{238}U$ ratio of U500 was measured on two ion counters over 6 replicate analyses of 200 seconds at 5 x 10^4 cps ^{238}U. The results are shown in Figure 6, with the mean of the measured data (blue) and the certificate value (grey) indicated. No mass fractionation correction has been made. No drift was observed, the gains are stable.

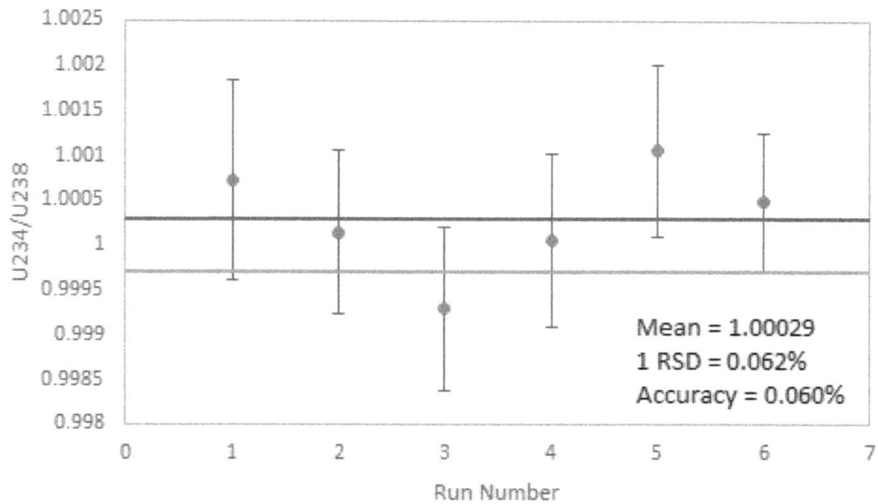

Figure 6 *No gain drift over the measured period, with external reproducibility of 0.06% 1RSD.*

6 ARE THEY STABLE RELATIVE TO FARADAYS?

Ion counter stability relative to a Zeptona Faraday was measured over 100 minutes. A 1.6 x 10^4 cps ion beam was measured for 2 s on ion counters and 10 s on the Faraday. A precision of ~0.4% 1 RSE was seen for each ion counters. An example is shown in Figure 7, showing no evidence of ion counter gain drift versus Faradays.

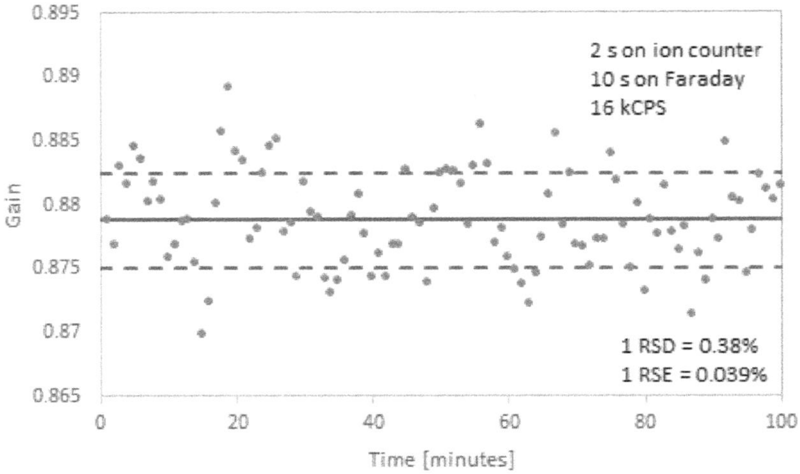

Figure 7 *Excellent stability over the 100 minutes of measurement, with no evidence of ion counter gain drift.*

7 WHAT'S A ZEPTONA FARADAY?

The Zeptona Faraday detector employs a single channel ATONA™ amplifier and rear Faraday detector optimised for minimal noise. Figure 8 shows the noise characteristics over a range of integration times, with ATONA™ amplifier data and theoretical resistor amplifier performance shown for reference. Zeptona achieves noise levels 5 times lower than those of the ATONA™ at all integration times.

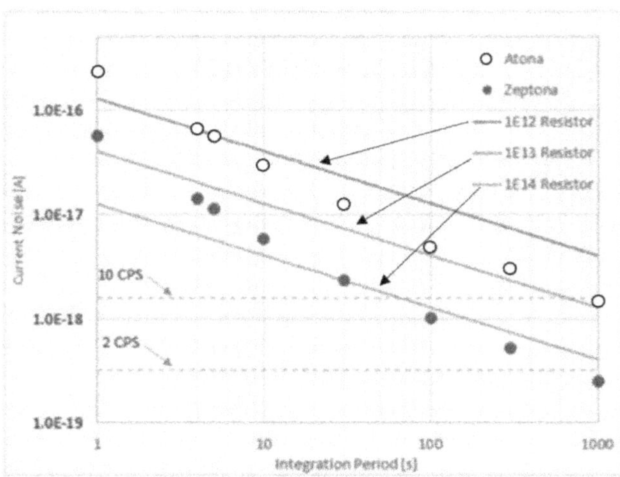

Figure 8 *Noise characteristics of the Zeptona amplifier over a range of integration times.*

8 WHAT IS THE ZEPTONA FARADAY NOISE LEVEL?

The Zeptona noise is comparable to the theoretical noise of a 1x 10^{14} ohm resistor at integration times of >10 seconds. The Zeptona Faraday was used to calibrate ion counter gain with very small ion signals. A 5 x 10^3 cps ^{238}U was measured on all ion counters (2 s integrations) and the Zeptona Faraday (10 s integrations). Figure 9 shows the results from a typical channel, with a precision of better than 0.15% (1 RSE) achieved after 50 cycles. This demonstrates that even at very low beam intensities the Faraday noise does not limit gain calibration precision.

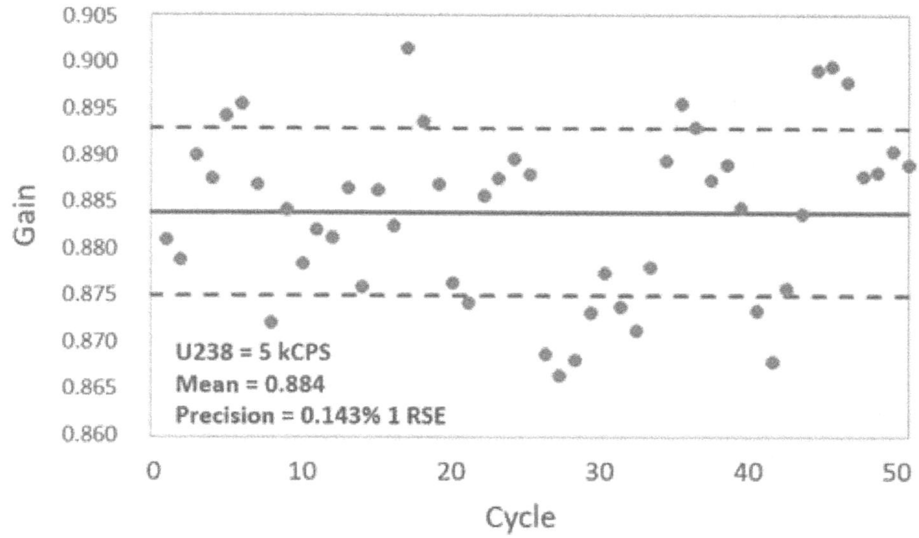

Figure 9 *Results from a typical channel, with a precision of better than 0.15% (1 RSE) achieved after 50 cycles.*

9 CONCLUSIONS

Our flexible, powerful multiple ion counting system offers the lowest noise and largest dynamic range:

- Up to 12 movable ion counters.
- Fixed Zeptona Faraday, optional Daly detector.
- Ion counting noise of 0.5 CPM, and single channel Faraday noise comparable to 1 x 10^{14} Ohm resistor amplifier.
- Excellent peak shape and flat over the focal plane – static and multi-dynamic ion counting analysis possible.
- Stable to < 0.3% RSD over 1 hour.
- Linear to < 0.3% RSD up to 300 kcps.
- Zeptona Faraday allows robust ion counter gain calibration with beams as small as a few thousand cps.
- 11 orders of magnitude dynamic range.

CHAPTER 15 AN INVESTIGATION INTO THE ROLE OF C-TYPE CYTOCHROMES AND EXTRACELLULAR FLAVINS IN THE BIOREDUCTION OF URANYL(VI) BY *SHEWANELLA ONEIDENSIS* USING FLUORESCENCE SPECTROSCOPY AND MICROSCOPY

D. L. Jones[a], M. B. Andrews[a], S. W. Botchway[b], A. Ward[b], J. R. Lloyd[c*] and L. S. Natrajan[a*]

[a]Department of Chemistry, School of Natural Sciences, The University of Manchester, Oxford Road, Manchester, M13 9PL, UK, [b] Central Laser Facility, Rutherford Appleton Laboratory, Harwell, Oxfordshire, OX11, 0QX, UK, [c]Department of Earth, Atmospheric and Environmental Sciences, School of Natural Sciences, The University of Manchester, Oxford Road, Manchester, M13 9PL, UK

*louise.natrajan@manchester.ac.uk; jon.lloyd@manchester.ac.uk

1 INTRODUCTION

Many sites across the world have been dealing with legacy radioactive contamination issues, resulting in an increased effort to investigate the role of these contaminants within the environment.[1] With new proposed investments in new build nuclear power stations in the UK comes a responsibility to deal with legacy and future wastes and possible contamination issues to ensure the safety of generations to come. Additionally, historical containment breaches at different stages during the fuel cycle, along with weapons testing and accidental spillage, has led to elevated levels of uranium within the environment. Successful remediation of these sites is essential to the sustainability of the nuclear industry, of increasing importance to realise net zero carbon emissions by 2050.[2] The oxidation state of uranium under a given set of environmental conditions is key to understanding and therefore predicting the chemical behaviour of the uranium in surface and near surface disposal in the form of uranyl(VI) which a linear di-oxo species, $U^{VI}O_2^{2+}$.[3] This species is water soluble as it predominately forms discrete monomeric and dimeric compounds including hydroxide and carbonate species, and as such can easily migrate within the environment from the original contamination site. Therefore, the development of remediation techniques whereby uranyl(VI) is reduced to its lower oxidation state form, U^{IV} are highly sought. This oxidation state is generally much less water soluble and most commonly precipitates from solution in anoxic conditions as the black mineral uraninite, $U^{IV}O_2$. In the case of enzymatic biotic reduction strategies, where common sub-surface and anoxic metal-reducing microbial communities (e.g. sulfate and iron reducers) can be exploited to preferentially biotically reduce $U^{VI}O_2^{2+}$ to $U^{IV}O_2$,[4,5] the presence of phosphorous from the microbial biomass can additionally result in the production of more soluble non uraninite U(IV) species.[5] These are generally poorly crystalline phases, often dubbed 'monomeric U(IV)', which can then be subjected to further chemical reactions including oxidation, further contributing to the higher than expected solubility of such species. It is therefore important to develop a fundamental understanding of the mechanisms of bio-reduction at the cellular level. *In situ* studies at the Rifle site in Colorado have shown that in the presence of acetate as an electron donor, naturally occurring bacteria dominated by *Geobacter genus*, are able to reduce uranyl(VI) leading to reduced concentrations of soluble uranium within the subsurface.[6] *Geobacter* species, alongside another widely studied bacterial *genus* in this field, *Shewanella*, are Gram-negative anaerobic bacteria.[7] Gram-negative bacteria possess both an inner and outer membrane, which in metal-reducers, have been shown to contain an abundance of *c*-type cytochromes. By contrast, Gram-positive bacteria lack an outer membrane but possess a thicker cell wall.[8]

Specifically, *Shewanella oneidensis* MR-1 has been widely studied in this field to investigate the role of cytochromes during the bioreduction process.[9,10,11] Organic electron donors such as lactate act as electron donors that feed the electron transport chain within the bacterium metabolic processes.[12,13] There are four mechanisms that have been proposed that result in electron transfer from the cell surface to an extracellular acceptor in *Shewanella oneidensis* MR-1. Of, these, two involve direct electron transport resulting from direct contact between the bacterium and the mineral or *via* nanowires, whereas the alternative two mechanisms invoke the involvement of soluble electron shuttles to mediate the electron transfer processes including secreted flavins or bound metal ions (here, uranium).[16] A total of 41 putative c-type cytochromes have been identified during the genome sequencing of *Shewanella oneidensis,* many of which include multi-heme-containing proteins that are postulated to be instrumental in the transfer of electrons within the electron transport chain.[14] Four specific cytochromes that are hypothesised to be heavily involved in the electron transport chain and therefore the respiratory mechanism of *Shewanella oneidensis* are MtrA, MtrB, MtrC and OmcA.[15] MtrA serves as an entry point for electron flow, MtrB relays electrons through the periplasm to the outer membrane and MtrC interacts with OmcA forming stable complexes as part of the electron transport chain.[15] Exposure of OmcA on the surface of the cell then enables interaction with electron acceptors, here, uranyl(VI).[15] Moreover, *Shewanella* species are known to secrete redox active flavin molecules. Flavin adenine dinucleotide (FAD) is known to be retained within the cell, whereas flavin mononucleotide (FMN) (and riboflavin) are secreted extracellularly, where FMN can catalytically mediate electron transport pathways during microbial uranyl(VI) reduction.[16] It has also been suggested that flavins likely interact, either directly or indirectly, with extracellular outer membrane cytochromes on the cell surface.

Microscopic and spectroscopic techniques used to study the reduction of uranium and other metals include X-ray based techniques such as X-ray absorption near edge structure (XANES) or extended X-ray absorption fine structure (EXAFS).[17] These techniques can involve complicated and often time consuming sample preparation and data analysis, and generally require millimolar sample concentrations on pre-prepared samples. Moreover, they are generally limited to the study of fixed cells. The use of complementary techniques that operate on the sub-micron scale are therefore attractive to study the mechanisms of bioreduction and the specific interactions of *Shewanella* cells with uranyl(VI) during the biosorption/bioreduction processes. We previously reported that fluorescence spectroscopy and microscopy can be used as a complementary technique both in the bulk and on the microscopic level to probe the speciation and lifetimes of *Geobacter sulfurreducens* bio-reducing uranyl(VI) samples *in situ*.[18] By monitoring the concentration of uranyl(VI) over a 24 hour period, a clear decrease followed by a sharp increase in soluble uranyl(VI) concentration, depicted by an undulating relationship of uranyl(VI) fluorescence intensity *vs* time suggested the formation of the one electron reduced species uranyl(V), followed by disproportionation to uranyl(VI) and uranium(IV).[18,19] Fluorescence lifetime imaging microscopy demonstrated, for the first time, a direct interaction between uranyl(VI) in solution and bacterial cells with approximately a 1000 fold decrease in the luminescence lifetime of the uranyl(VI) emission observed upon contact with the surface of the bacteria.[18] This measurable emission lifetime decrease is a direct result of quenching of the charge separated excited state of the ligand-to metal charge transfer (LMCT) electronic transition of uranyl(VI) as a result of electron transfer (rapid charge recombination) processes with proximate redox active partners, here in cyctochromes and flavins. This demonstrates the ability of fluorescence microscopy to probe the cell surface redox interactions of bacteria and uranyl(VI), with the possibility of directly observing the effect of outer membrane *c*-type cytochromes and other biological structures in electron transfer to uranyl(VI). More recently, we provided direct evidence of the presence of a pentavalent uranyl(V) species as a major product in the post reduction solid phases of *Shewanella oneidensis* MR-1 (up to 59 % after 4.5 hours)[11] and that uranyl(V) may be stabilized for up to 120 hours (with 26–28% U(V) still present with the biomass. The existence of the uranium(V) species is much more prolonged than originally predicted and provides evidence that uranyl can remain associated with the biomass during the one electron bioreduction and disproportionation process.

Here, we report an investigation into the mechanisms of the enzymatic bioreduction of uranium, from $U^{VI}O_2^{2+}$ to insoluble U^{IV} by *Shewanella oneidensis* by the use of luminescence spectroscopy and confocal fluorescence microscopy combined with fluorescence lifetime image mapping. The use of genetically engineered deletion mutants of the *Shewanella oneidensis* MR-1 strain, lacking outer membrane cytochromes, both inner and outer membrane cytochromes and lacking in the ability to secrete flavin co-factors, has provided a much-needed deeper insight into the role of *c*-type cytochromes in live resting state microbial communities within the bacterial electron transport chain and the effect of extracellular redox active flavins on the overall efficacy of the bioreduction process.

2 EXPERIMENTAL PROCEDURES

2.1 Maintenance and growth of organisms

Four strains of *S. oneidensis* were donated by Jeffrey Gralnick (University of Minnesota); JG 274 – wild type, JG 596 – lacking outer membrane (OM) cytochromes, JG 1453 – lacking inner membrane (IM) and outer membrane (OM) cytochromes, JG 1758 – lacking the ability to secrete extracellular flavins including flavin mononucleotide (FMN), were grown at 30°C under aerobic conditions on LB agar plates for 24 hours before being transferred to LB broth liquid medium for a further 24 hours.

2.2 Resting cell experiments

All manipulations of cells were carried out under an atmosphere of N_2. Late-log-phase cultures were harvested by centrifugation and washed twice in $NaHCO_3$ buffer (30 mM, pH 7, degassed with N_2-CO_2 80:20 mix) according to literature precedent.[18] Aliquots of the washed cell suspensions were added to a final concentration of 0.5 mg mL-1 dry weight biomass, by use of a syringe fitted with a needle, to anaerobic bottles sealed with butyl rubber stoppers containing uranyl(VI) acetate (5 mM) in $NaHCO_3$ buffer (30 mM, pH 7) creating the microcosm for the experiment. Lactate (10 mM) was added to the cultures as the electron donor. Samples were removed every hour, by use of a syringe fitted with a needle and filtered into a low temperature quartz EPR tube or placed onto a quartz slide and sealed with a cover slip for low temperature luminescence (flash freezing in liquid nitrogen at 77 K) and microscopy studies respectively.

Shewanella oneidensis; JG 274 – wild type, JG 596 – lacking OM cytochromes, JG 1453 – IM and OM cytochromes, JG 1758 – lacking the ability to secrete extracellular flavins, were resuspended in 30 mM bicarbonate buffer supplemented with uranyl(VI) acetate (5 mM) and sodium lactate (10 mM) as an electron donor for the radionuclide reduction in a sealed serum bottle under a nitrogen atmosphere for sampling by luminescence spectroscopy and liquid scintillation counting (Perkin-Elmer Quantulus instrument) at regular intervals during the bioreduction process.

Table 1 *Summary of the differences between each of the four Shewanella oneidensis strains used during this investigation.*

***Shewanella oneidensis* Strain/deletion mutant**	**Summary**
JG 274	Wild type, no alterations
JG 596	Lacking outer membrane cytochromes
JG 1453	Lacking inner and outer membrane cytochromes
JG 1758	Lacking the ability to produce flavins

2.3 Optical spectroscopy and microscopy

Steady state emission spectra were recorded on an Edinburgh Instrument FP920 Phosphorescence Lifetime Spectrometer equipped with a 5 Watt microsecond pulsed xenon flash lamp and a red sensitive photomultiplier in peltier (air cooled) housing, (Hamamatsu R928P). Lifetime data were recorded following 420 nm excitation with a 5 Watt xenon flash lamp (Edinburgh Instruments), using multichannel scaling. All experiments were repeated in triplicate with the same excitation wavelength and excitation and emission monochromator slit widths and corrected for the excitation source and detector response using the F900 software in built correction files. The one photon-excitation FLIM-PLIM used is a commercial system from Becker and Hickl with a 405 nm pulsed laser system. Fluorescence lifetime images were obtained using a modified confocal microscopy apparatus, constructed in the Central Laser Facility, which has a Nikon eC1 scan head. Laser light was focused through a x 60 water immersion objective with an NA of 1.2 on an inverted Nikon microscope (TE2000-U). Steady state spectra were recorded using an Ocean Optics USB2000+ spectrometer.

3 RESULTS AND DISCUSSION

Luminescence spectroscopy of the uranyl(VI) ion has proven an invaluable spectroscopic technique in identifying it's solution and solid state chemical speciation. The emission process is now widely accepted to be a phosphorescent ligand-to-metal charge transfer (LMCT) process involving promotion of a bonding 2p electron of the axial oxygen molecules to a 5f orbital of non-bonding character.[20] This characteristic emission is typically centred around ca. 520 nm and up to six vibrational hot bands can be observed as a result of strong coupling of the excited electronic state with the total symmetric vibrational mode of the uranyl moeity.[21] Long luminescence lifetimes on the order of microseconds are typical of uranyl(VI) salts and compounds, characteristic of the long lived excited state.[21] The combination of luminescence spectroscopy (fluorescence and phosphorescence) with optical microscopy and lifetime image mapping (in fluorescence and phosphorescence mode) enables optical images to be obtained of uranyl(VI) species itself without the use of fluorescent dye staining molecules in contrast to the majority of optical imaging protocols utilized in the biological sciences. Furthermore, the changes in uranyl(VI) lifetime values with lifetime image mapping can be utilized to directly observe uranyl(VI)-bacterial interactions *in situ* during sorption and reduction processes in addition to highlighting uranyl(VI) localisation at the microbial cell surface.[22]

In this work, the emission of uranyl(VI) acetate remaining in solution following addition to the microbial cultures of *Shewanella oneidensis* MR-1 after centrifugation and removal of the biomass over a 24 hour period was analysed at room temperature. The spectra show broad featureless spectra typical of a range of uranyl(VI) solution species.[23,24] However, in frozen solution at 77K the spectra are well resolved as a direct result of reduced luminescence quenching effects at low temperature, greatly simplifying the bulk luminescence analysis. As such, the total emission of all uranyl(VI) remaining in solution over a 24 hour period was analysed at 77K. Additionally, the luminescence lifetimes of soluble uranyl(VI) remaining in solution recorded throughout the bioreduction experiments for all *Shewanella* strains are all similar, and follow mono-exponential decay kinetics suggesting the presence of a single emissive uranyl(VI) species on the microsecond timescale at 77K.[18] The recorded lifetimes are: 1040.60 ± 1.059 μs (JG 274 – wild type), 1024.56 ± 0.948 μs (JG1758 – lacking ability to secrete extracellular flavins), 974.52 μs ± 1.149 (JG 596 – lacking OM cytochromes) and 1000.70 μs ± 1.115 (JG 1453 – lacking IM and OM cytochromes).

The concentration of uranyl(VI) was next monitored using luminescence spectroscopy over the course of a bioreduction experiment for each *S. oneidensis* bacterial strain. At hourly time points for the initial 8 hours, where the majority of the bioreduction is known to occur, and then again at 24 hours; an aliquot of the solution was removed from the reduction microcosm and frozen in liquid nitrogen where a fluorescence spectrum was obtained under standardized instrumental set up. A small amount of the supernatant, removed via a syringe filter to remove any biomass, was also analysed to determine the total uranium concentration by liquid scintillation counting and ICP-MS analysis. The data reported are an average of three independent bioreduction experiments for each bacterial strain (Figure 1). For the JG274 (wild type) strain and the JG1758 (lacking ability to secrete extracellular flavins) strain, the fluorescence intensity of uranyl(VI) showed a general decrease and after 24 hours and 48 hours respectively solutions were non emissive, suggesting that the microbial reduction had proceeded to completion. This observation was supported by liquid scintillation counting which showed a general linear decrease in soluble uranyl(VI) concentration over the same time period (Figure 1).

Previous luminescence data recorded during the bioreduction of uranyl(VI) acetate with *Geobacter sulfurreducens* highlighted an undulating intensity pattern of total uranyl(VI) concentration versus time by luminescence spectroscopy.[18] The initial decrease in total emission intensity of the uranyl(VI) signal and subsequent sharp increase observed in hours 2-3 of the bioreduction process and further steady decrease in emission from hour 4 onwards, was attributed to the formation and relatively rapid disproportionation of a kinetically unstable metastable uranyl(V) ($U^{V}O_2^{+}$) intermediate which is non-emissive under the experimental conditions employed. This same effect is observed during the reduction with *Shewanella oneidensis*, with an undulating pattern obtained reproducibly for JG274 which is attributed to the formation of an unstable $U^{V}O_2^{+}$ intermediate that rapidly disproportionates to form the observed $U^{VI}O_2^{2+}$ emissive species alongside insoluble U^{IV} nanoparticles in agreement with previous data[11] (Figure 1). This observation strongly suggests that the bioreduction process in wild type *Shewanella oneidensis* also involves one-electron reduction of uranyl(VI) followed by disproportionation, consistent with previous reports of *Shewanella oneidensis* MR-1[11] and *Geobacter sufurreducens*.[25]

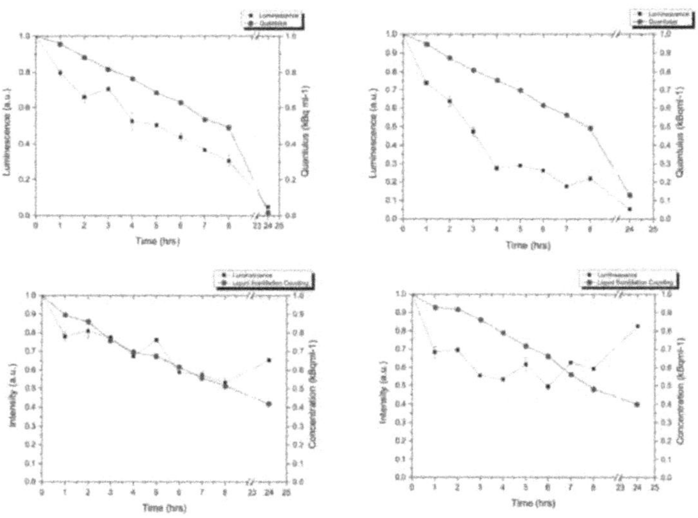

Figure 1 *Plots of luminescence intensity (squares) of uranyl(VI) normalised to time = 0 (excitation wavelength 280 nm, 77 K) and liquid scintillation counting data measured using a Quantulus instrument (circles) for Shewanella oneidensis JG274 – wild type (top left) and Shewanella oneidensis JG1758 – lacking ability to secrete extracellular flavins (top right), Shewanella oneidensis JG596 – lacking OM cytochromes (bottom left), Shewanella oneidensis JG1453 – lacking inner membrane and outer membrane cytochromes (bottom right), over the course of a 24 hour bioreduction with uranyl(VI) acetate (5 mM) and lactate (10 mM) as the electron donor.*

It is clear in Figure 1 that the decrease in uranyl(VI) emission in the JG1758 strain occurs at a faster initial rate, with reaction rate values of 0.08 intensity hr^{-1} for JG274 and 0.21 intensity hr^{-1} for JG1758. The presence of flavins may be affecting the obtained data since flavins are hypothesised to quench uranyl(VI) emission due to spectral overlap of the respective absorption spectra and resultant energy transfer from the excited state of uranyl(VI).[26] This effect is also observed during the reduction experiments when the luminescence intensity at time = 0 is compared to a uranyl(VI) acetate emission spectrum which has not been in contact with bacteria. The uranyl(VI) emission of the JG274 (wild type) decreases dramatically in the presence of the bacteria which are secreting flavins, whereas a the smaller decrease in emission intensity is observed for the JG1758 strain which is not secreting a significant concentration of flavins and therefore not quenching the uranyl(VI) emission. A titration carried out with oxidised flavin mononucleotide (FMN) shown in Figure 2 demonstrates the complete quenching of uranyl(VI) emission following the addition of 0.08 equivalents of FMN. Flavins are well known in biological process to be 2-electron reducing agents[27] but as both the flavin and uranyl(VI) are in their oxidised forms it seems unlikely that reduction is occurring and the decrease in emission is more likely to be due to the quenching effect induced by spectral overlap of the FMN absorption and the uranyl(VI) emission through energy transfer.

When following the uranyl(VI) emission signal for the JG596 (lacking OM cytochromes) and JG1453 (lacking IM and OM cytochromes) strains (Figure 3) an initial decrease in uranyl(VI) concentration is observed which could be due to transient sorption of the uranyl(VI) onto the bacterial cells, or more likely an indication that some residual reduction activity with this mutant remains albeit significantly reduced compared with the wild type. This observation alongside the decrease in total uranium concentration over time as demonstrated by the liquid scintillation data, suggests that significant sorption to the numerous available sorption sites on the bacterium surface occurs as expected but that bioreduction is impeded. Notably, over a 24 hour period there is little decrease in the emission intensity of uranyl(VI) which suggests that the cytochromes on both the inner and outer membrane of the bacteria play an important role in the electron transport chain during the reduction process. This can be seen by the intensity profile of the uranyl(VI) emission at 24 hours, which indicates the presence of uranyl(VI) in high concentrations (relative emission intensities of 0.65 a.u. and 0.82 a.u. respectively) compared to the reductions by JG274 and JG1758, where most of the uranyl(VI) has been reduced within 24 hours.

To investigate whether the cytochromes are capable of reducing uranyl(VI) directly, titrations were carried out with pure reduced cytochrome *c* from equine heart (Figure 2). Over the course of the titration, the intensity of the uranyl(VI) emission observed decreased in experiments, which was complemented by an increase a broad emission band at ca. 350 nm. This may be due to the presence of uranium(IV) species that has been formed by reduction of the uranyl(VI) by the heme centres of cytochrome *c*; a broad emission band for the +IV oxidation state of uranium has been reported in the literature at 365 nm.[28] Solution lifetime data could not be collected available due to the short lived nature of the band at ca. 350 nm, which was less than the instrument response function of the 5W microsecond flashlamp employed (< 5 μs) and the fact that this ultra-violet emission was too weak in intensity to be measured using available mW UV LED lasers.

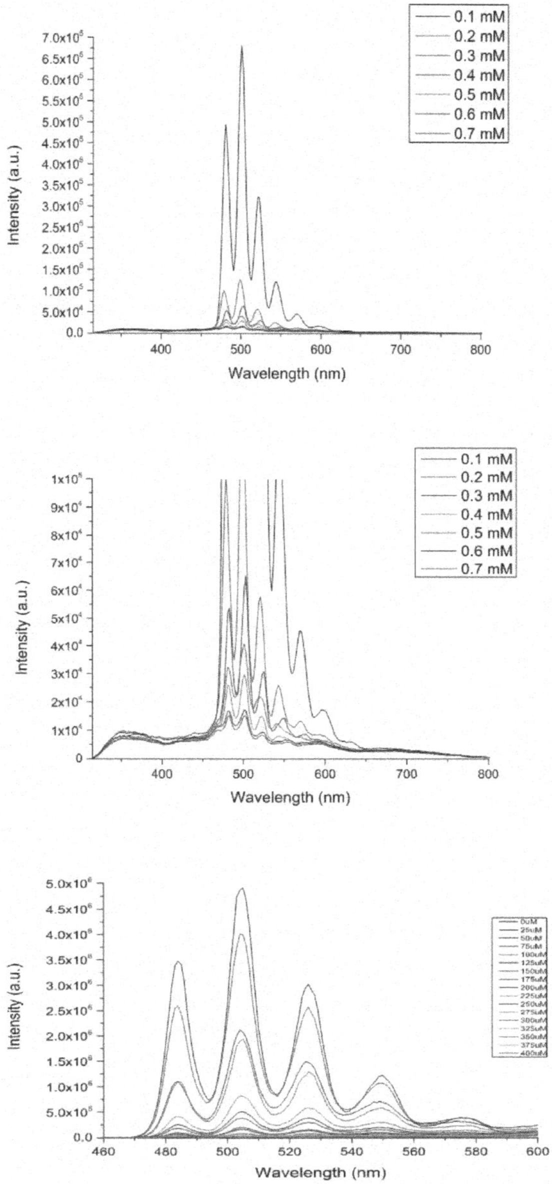

Figure 2 *Steady state luminescence spectra showing the decrease in uranyl(VI) acetate emission in water centred at 510 nm (top) and the possible increase in uranium(IV) emission at 350 nm (middle) with increasing addition of cytochrome c (0 – 7 mM) and bottom, with increasing addition of FMN (0-400 μM), following excitation at 280 nm, 77 K.*

We next turned to the use of confocal microscopy and fluorescence and phosphorescence lifetime image mapping to probe the distribution of uranyl(VI) within the isolated microbial biomass at sub-micron spatial resolution. We have previously shown that optical microscopy techniques commonly employed in the biological sciences such as Fluorescence Lifetime Image Microscopy (FLIM) and Phosphorescence Lifetime Image Microscopy (PLIM) can be used to temporally and spatially resolve the lifetimes of species present with sub-micron resolution.[22] This permits *in situ* images to be taken during a bioreduction process without fixing of the cells, enabling more information to be gained on the interactions occurring between the uranium and the cell surface. The experimental procedure was carried out as described previously[22] with aliquots of the centrifuged and washed biomass in buffer being placed on a glass slide and experiments performed at room temperature.

Little or no auto-fluorescence was observed from all the bacterial strains upon direct excitation into the uranyl(VI) LMCT absorption band (to avoid coincidental excitation of biological chromophores) with a 405 nm laser in the one-photon confocal experiment in the absence of uranyl(VI). However, upon the addition of uranyl(VI) JG274 (wild type) and JG1758 (lacking ability to secrete extracellular flavins), which have both been shown to reduce $U^{VI}O_2^{2+}$ in work presented above, demonstrated variable fluorescence lifetimes across the cell surface of the bacteria at the start of the bioreduction (time = 0). Images are shown in Figure 3 (at time = 2 hours) and lifetime distributions are presented in Figure 4. The uranyl(VI) fluorescence lifetimes across the surface of the JG274 (wild type) and JG1758 (lacking ability to excrete extracellular flavins) cells show shorter and relatively narrow fluorescence temporal distributions across the FLIM images of ca. 2.2 - 3.0 ns and 2.7- 4.4 ns respectively (Figure 4). The fact that the uranyl(VI) emission is variable across the bacterial cells at the very beginning of the bioreduction process suggests that uranyl(VI) is interacting strongly with different chemical species on the cell surface since the emission is undergoing a competitive quenching process.[18] The quenching of uranyl(VI) emission occurs when a non-radiative pathway for electron relaxation is present in the sample. In principle, this could occur due to changes in solvent speciation, changes in complexation or vibrational quenching and/or spectral overlap of the uranyl(VI) excited state with that of another species (here, cytochromes/flavins).[3] In this case, since vibrational quenching is expected to be time averaged, these observations are most likely to be directly related to the interactions between the uranyl(VI) and a surface bound electron donating species on the bacterial cell.[18] In addition, it is well known that quenching of uranyl(VI) emission *via* electron transfer with Fe(II) ions is extremely effective (such that no emission from the uranyl(VI) ion can be recorded even at cryogenic temperatures).

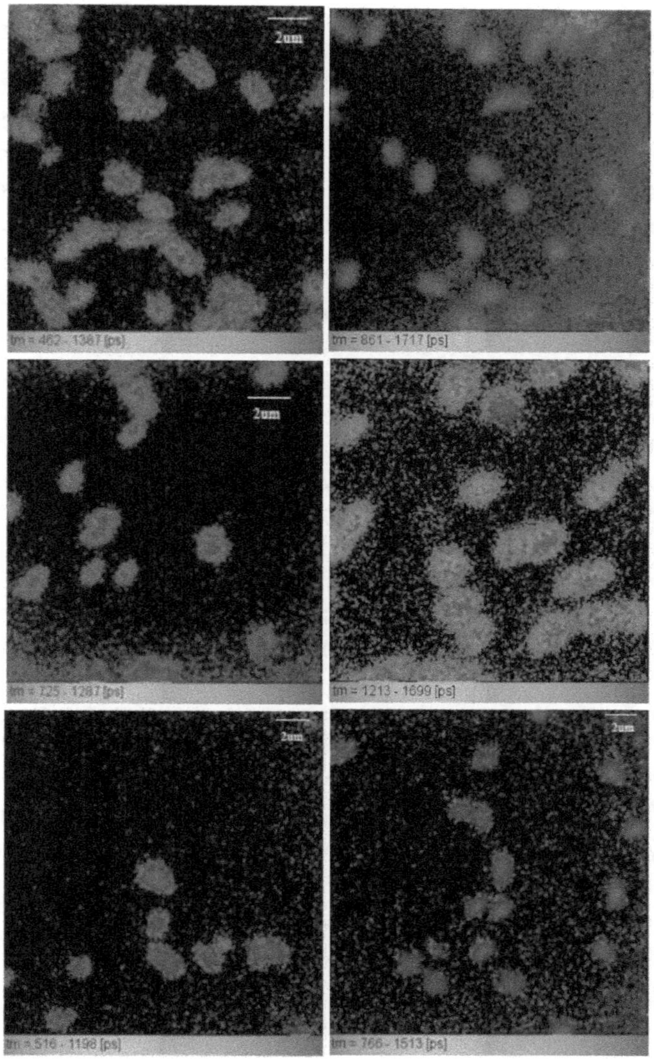

Figure 3 *Representative one-photon fluorescence lifetime image microscopy map of lifetime data for S. oneidensis JG274 – wild type (top row) and JG1758 – lacking ability to secrete extracellular flavins (middle row), JG 596- lacking outer membrane cytochromes (bottom left) and JG 1453- lacking inner and outer membrane cytochromes (bottom right) at T=2 during a 24 hour reduction with uranyl(VI) acetate (5 mM) and lactate (10 mM) as the electron donor. Excitation wavelength = 405 nm, 298 K.*

Here, excited state quenching of the uranyl(VI) emission is substantially more pronounced in the presence of both inner and outer membrane cyctochromes, manifested by much narrower temporal distributions in the uranyl(VI) emission. Notably, there was negligible phosphorescence signal in the time gated PLIM window (μs time domain), indicative that no emissive uranyl(VI) salts or extracellular uranyl(VI) species are present as we previously observed in the case of *Geobacter sufurreducens* and that all uranyl(VI) is associated with the bacterial cells

Intriguingly, the same variation in fluorescence lifetimes were not observed in JG596 (lacking OM cytochromes) and JG1453 (lacking IM and OM cytochromes), Figure 4. These are the strains which do not appreciably reduce $U^{VI}O_2^{2+}$, indicating that the redox partners responsible for the quenching of the uranyl(VI) emission are not present in these deletion mutants, which suggests it may be related to the absence of key *c*-type cytochromes in these mutants. The fluorescence lifetime distributions of these deletion strains JG596 (lacking OM cytochromes) and JG1453 (lacking IM and OM cytochromes) are significantly broader and the average lifetimes longer (range, 5.8 – 34 ns and 8.7 – 11.8 ns, Figure 4). The absence of both inner and outer membrane cytochromes in the JG596 and JG1453 mutant strains also results in an inability to reduce $U^{VI}O_2^{2+}$ in agreement with the steady state fluorescence spectral data presented in Figure 1. This suggests that the cytochromes play an important role during the electron transfer pathway in the bioreduction of uranyl(VI), directly interacting with the uranyl(VI) on the surface of the bacteria causing the quenching of the emission observed in the samples where uranyl(VI) is successfully reduced. Together, these data strongly suggest that the absence of cytochromes inhibits electron transfer processes that reduce uranyl(VI) in all mutant strains and that the involvement of both inner and outer membrane cytochromes in the bioreduction process has a more substantial effect than that of the secreted flavins.

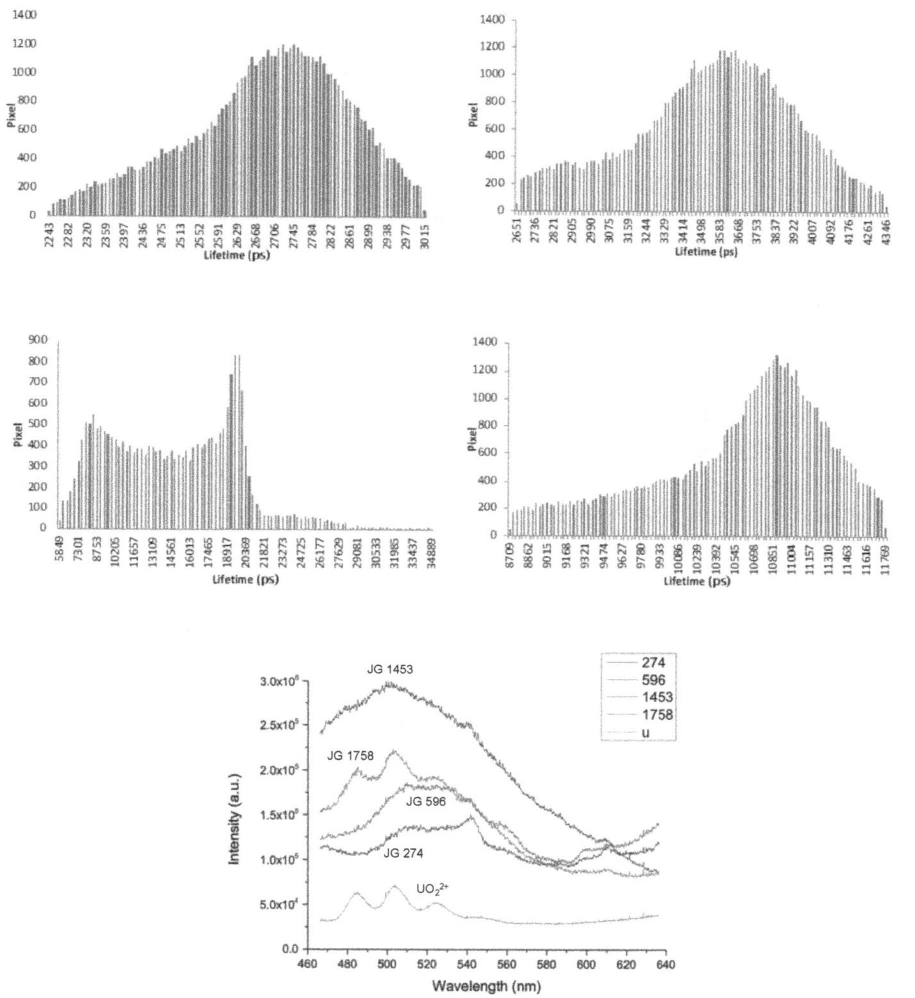

Figure 4 *Representative lifetime histograms across a FLIM image for S. oneidensis JG 274 – wild type (top left and JG1758 – lacking ability to secrete extracellular flavins (top right), JG 596- lacking outer membrane cytochromes (bottom left) and JG 1453- lacking inner and outer membrane cytochromes (bottom right) at T=2 during a 24 hour reduction with uranyl(VI) acetate (5 mM) and lactate (10 mM) as the electron donor. Excitation wavelength = 405 nm, 298 K. Below: representative steady state emission spectra collected during the FLIM experiments for all bacterial strains compared to uranyl in buffer (u), excitation wavelength = 405 nm, 298 K showing a variation in emission profile due to differing interactions of uranyl(VI) with the bacteria.*

It is worth noting here that in addition to temporal variations in the FLIM data, the fluorescence intensities of the confocal microscopy images of the JG 596 and JG 1453 bacterial strains showed relatively lower emission intensities when compared to those of the flavin secretion deletion mutant (JG 1758) and the wild type *S. oneidensis* MR-1 (JG 274) recorded under identical experimental conditions. This observation may be explained by a reduced concentration of uranyl(VI) emitting centres that are sorbed onto the surface of the bacteria due to the lack of interaction with *c*-type cytochromes. This hypothesis is supported by the fact that the JG596 strain (lacking OM cytochromes) in particular was not as emissive as the JG274 (wild type) strain. The most emissive sample was the JG1758 (lacking ability to secrete extracellular flavins) strain. Flavin cofactors, in this case, FMN are hypothesised to quench uranyl(VI) emission by resonant energy transfer by spectral matching (in their oxidised forms) and *via* electron transfer processes in the reduced forms and therefore when their secretion by the bacteria is hindered they no longer quench the emission. This quenching effect suggests there is a direct interaction between the secreted flavin molecules and the uranyl(VI) species which influences the bioreduction mechanism of uranyl(VI) ions on and/or proximate to the bacterial cell surface.[31]

4 CONCLUSIONS

In summary, we have examined the ability of *Shewanella oneidensis MR-1* and three deletion mutants to reduce $U^{VI}O_2^{2+}$ over a 24 hour period with the use of steady state and time resolved luminescence spectroscopy, alongside one-photon fluorescence and phosphorescence confocal microscopy and lifetime image mapping to visualize the enzymatic reduction processes in live cells *in situ*. Experiments with three deletion mutant strains of *Shewanella oneidensis* have provided deeper insight into the role of the inner and outer membrane *c*-type cytochromes and the secreted extracellular flavin mononucleotide, FMN; JG274 (wild type), JG596 (lacking OM cytochromes), JG1453 (lacking IM and OM cytochromes) and JG1758 (lacking ability to secrete extracellular flavins). Steady state luminescence spectroscopy of the supernatants during the bioreduction processes by these strains have shown that those lacking in cytochromes, JG596 (OM) and JG1453 (IM and OM), do not reduce the $U^{VI}O_2^{2+}$ over the same time period as those which do, and that differences in inner and outer membrane cytochromes affects the rate at which enzymatic reduction occurs.

Both fluorescence and phosphorescence microscopy and lifetime imaging mapping has suggested that the presence of the cytochromes on the bacterial surface has a significant quenching effect on the luminescence lifetime of any uranyl(VI) present, observed by the lack of substantial temporal variation in lifetimes across the bacterial surface when the inner or outer cytochromes are not present. This optical 'redox map' suggests that the cytochromes are directly interacting with the uranyl(VI) on the surface of the bacteria, indicating that the reduction must be a direct process between the two species. These data additionally suggest that bound flavins may be disruptive to this process. This confirms that both inner and outer membrane *c*-type cytochromes play an crucial role in the electron transport chain during the bioreduction of uranyl(VI), and that outer membrane cyctochromes in particular may be more closely associated with uranyl(VI) ions than previously considered.

5 ACKNOWLEDGEMENTS

We are grateful to the EPSRC for funding a Nuclear First DTC studentship (DJ), a Career Acceleration Fellowship (LN) and postdoctoral funding (MA), Grant EP/G004846/1, the Leverhulme Trust for additional postdoctoral funding (MA) and a research Leadership award (LN), Grant RL-2012-072, and the STFC for funding direct access to the Laser for Science Facilities at Rutherford Appleton Laboratories and for support through the STFC funded Environmental Radioactivity Network (Env-Rad-Net). We also thank Professor Jeffrey Gralnick from the University of Minnesota for provision of the *Shewanella oneidensis* deletion mutants.

6 REFERENCES

1. F.R. Livens, M.J. Jones, A.J. Hynes, J.M. Charnock, J.F.W Mosselmans, C. Hennig, H. Steele, D. Collison, D.J. Vaughan, R.A.D. Pattrick, W.A. Reed, L.N. Moyes, *J. Environ. Radioact.* 2004, **74**, 211.
2. S. Chakraborty, F. Favre, D. Banerjee, A.C. Scheinost, M. Mullet, J.J. Ehrhardt, J. Brendle, L. Vidal, L. Charlet, *Environ. Sci Technol.*, 2010, **44**, 3779.
3. L.S. Natrajan, *Coord. Chem. Rev.* 2012, **256**, 1583.
4. J. Janeczek, R.C. Ewing, *J. Nucl. Mater.* 1992, **190**, 128; W. You, W. Peng, Z. Tian, M. Zheng, *Sci. Tot. Environ.*, 2021, **798**, 149107.
5. J.D. Wall, L.R. Krumholz, *Annu. Rev. Microbiol.* 2006, 60; D.S. Alessi, B. Uster, H. Veeramani, E.I. Suvorova, J.S. Lezama-Pacheco, J.E. Stubbs, J.R. Bargar, R. Bernier-Latmani, *Environ. Sci. Technol.*, 2012, **46**, 6150; R. Bernier-Latmani, H. Veeramani, E. D. Vecchia, P. Junier, J.S. Lezama-Pacheco, J.O Sharp, N.S. Wigginton, J.R. Bargar, *Environ. Sci. Technol.*, 2010, **44**, 9456.
6. J.R. Bargar, K.H. Williams, K.M. Campbell, P.E. Long, J.E. Stubbs, E.I. Suvorova, J.S. Lezama-Pacheco, D.S. Alessi, M. Stylo, S.M. Webb, J.A. Davis, D.E. Giammar, L.Y. Blue, R. Bernier-Latmani, *Proc. Natl. Acad. Sci. U.S.A.* 2013, **110**, 4506.
7. J.R. Lloyd, J. Chesnes, S. Glasauer, D.J. Bunker, F.R. Livens, D.R. Lovley, *Geomicrobiol. J.* 2002, **19**, 103.
8. L.S. Keith, O.M. Faroon, B.A. Fowler, Uranium; Elsevier Academic Press Inc: San Diego, 2007.
9. G. Reguera, *Int. Microbiol.* 2015, **18**, 151.
10. Y. Yang, S. Wang, T.E. Albrecht-Schmitt, *Chem. Geol.* 2014, **387**, 59; A.S. Beliaev, D.A. Saffarini, J.L. McLaughlin, D. Hunnicutt, *Mol. Microbiol.* 2001, **39**, 722.

11. G.F. Vettese, K. Morris, L.S. Natrajan, S. Shaw, T. Vitova, Jurij Galanzew, D.L. Jones, J.R. Lloyd, *Environ. Sci. Technol.* 2020, **54**, 2268.
12. D.R. Lovley, E.J.P. Phillips, *Appl. Environ. Microbiol.* 1988, **54**, 1472.
13. D.R. Lovley, E.J.P. Phillips, D.J. Lonergan, *Appl. Environ. Microbiol.* 1989, **55**, 700.
14. M. Breuer, K.M. Rosso, J. Blumberger, J.N. Butt, *J. R. Soc. Interface*, 2015, 12; H.C.Gao, S. Barua, Y.L. Liang, L. Wu, Y.Y. Dong, S. Reed, J.R. Chen, D. Culley, D. Kennedy, Y.F. Yang, Z.L. He, K.H. Nealson, J.K. Fredrickson, J.M. Tiedje, M. Romine, J.Z. Zhou, *Microb. Biotechnol.* 2010, **3**, 455.
15. S.K. Tai, G. Wu, S.S. Yuan, K.C. Li, *BMC Genomics* 2010, 11.
16. Y. Suzuki, Y. Kitatsuji, T. Ohnuki, S. Tsujimua, *Phys. Chem.* 2010, **12**, 10081; M.J. Edwards, G.F White, M. Norman, A. Tome-Fernandez, E. Ainsworth, L. Shi, J.K Fredrickson, J.M Zachara, J.N Butt, D.J Richardson, T.A Clarke, *Sci. Rep.*, 2015, **5**, 11677.
17. S. Kushwaha, B. Sreedhar, P. Padmaja, Langmuir, 2012, **28**, 16038.
18. D.L. Jones, M.B. Andrews, A.N. Swinburne, S.W. Botchway, A.D. Ward, J.R. Lloyd, L.S. Natrajan, *Chem. Sci.* 2015, **6**, 5133.
19. J.C. Renshaw, L.J.C. Butchins, F.R. Livens, I. May, J.M. Charnock, J.R. Lloyd, *Environmental Science & Technology*, 2005, **39**, 5657.
20. G. Meinrath, *J. Radioanal. Nucl. Chem.* 1997, **224**, 119.
21. M.P. Redmond, S.M Cornet, S.D. Woodall, D. Whittaker, D. Collison, M. Helliwell, L.S. Natrajan, *Dalton Trans.* 2011, **40**, 3914.
22. E. Baggaley, S.W. Botchway, J.W. Haycock, H. Morris, I.V. Sazanovich, J.A.G. Williams, J.A. Weinstein, *Chem. Sci.* 2014, **5**, 879.
23. Z. Wang, J.M. Zachara, C. Liu, P.L. Gassman, A.R. Felmy, S.B. Clark, *Radiochim. Acta* 2008, **96**, 591.
24. W.D. Burgos, J.T. McDonough, J.M. Senko, G.X. Zhang, A.C. Dohnalkova, S.D. Kelly, Y. Gorby, K.M. Kemner, *Geochim. Cosmochim. Acta*, 2008, **72**, 4901.
25. N.K.W. Law, J.C. Renshaw, I. May, F.R. Livens, J.R. Lloyd, *Abstracts of the General Meeting of the American Society for Microbiology,* 2007, **107**, 569.
26. H. von Canstein, J. Ogawa, S. Shimizu, J.R. Lloyd, *Appl. Environ. Microbiol.* 2008, **74**, 615.
27. V. Balzani, L. Moggi, M.F. Manfrin, F. Bolletta, *Coord. Chem. Rev.,* 1975, **15**, 321.
28. G.F White, M. J. E., L. Gomez-Perez, D.J. Richardson, J.N. Butt, T.A. Clarke, *Advances in Microbial Physiology*, 2016, **68**, 87.
29. E. Hashem, A.N. Swinburne, C. Schulzke, R.C. Evans, J.A. Platts, A. Kerridge, L.S. Natrajan, R.J. Baker, *RSC Adv.* 2013, **3**, 4350.
30. T.J.D. Baker, J.R.G. Thorne, *Inorg. Chem. Commun.*, 1987, **26**, 1721.
31. D. Kim, H.G. Ryu, K.H. Ahn, *Org. Biomol. Chem.* 2014, **12**, 4550.

Isotope Index